The Plaston Concept

Isao Tanaka · Nobuhiro Tsuji · Haruyuki Inui
Editors

The Plaston Concept

Plastic Deformation in Structural Materials

 Springer

Editors
Isao Tanaka
Department of Materials Science
and Engineering
Kyoto University
Kyoto, Japan

Nobuhiro Tsuji
Department of Materials Science
and Engineering
Kyoto University
Kyoto, Japan

Haruyuki Inui
Department of Materials Science
and Engineering
Kyoto University
Kyoto, Japan

ISBN 978-981-16-7717-5   ISBN 978-981-16-7715-1   (eBook)
https://doi.org/10.1007/978-981-16-7715-1

This Springer imprint is published by the registered company Springer Nature Singapore Pte Ltd.
The registered company address is: 152 Beach Road, #21-01/04 Gateway East, Singapore 189721, Singapore

Preface

This book presents the novel concept of plaston, which accounts for the high ductility or large plastic deformation of emerging high-performance structural materials, including bulk nanostructured metals, hetero-nanostructured materials, metallic glasses, intermetallics, and ceramics. The plaston concept was first proposed within an intensive discussion and collaboration among members of the Elements Strategy Initiative for Structural Materials (ESISM), a Japanese national project funded by MEXT for the period of 2012–2021. This concept has been continuously discussed in the annual international symposia on "fundamental issues of structural materials" organized by ESISM and in symposia at the annual meeting of the Japan Institute of Metals. The authors of the chapters in this book have played active roles in those discussions.

The book consists of four parts. The first part is a single chapter and describes an introduction to the plaston concept and strategy to manage both high strength and large ductility in advanced structural materials, on the basis of unique mechanical properties of bulk nanostructured metals.

In the second part, simulations of plaston and plaston induced phenomena are reviewed. Chapter 2 shows free-energy-based atomistic study of nucleation kinetics and thermodynamics of defects in metals from the viewpoint of plastic strain carrier. Chapter 3 describes atomistic study of disclinations in nanostructured metals. Chapters 4 and 5 are dedicated to first principles calculations of collective motion of atoms through phonon calculations and dislocations through large-scale calculations.

The third part consists of five chapters. It presents the analyses based on various experiments on plaston. Chapter 6 explains the concept of plaston as an elemental deformation process involving cooperative atom motion. Chapter 7 shows the characterization of lattice defects associated with deformation and fracture in alumina by transmission electron microscopy (TEM). Chapter 8 describes nanomechanical characterization using nanoindentation technique and TEM. Chapter 9 shows a synchrotron x-ray study on plaston in metals. Chapter 10 focuses on microstructural crack tip plasticity, which controls the growth of small fatigue crack.

The fourth part is devoted to the design and development of high-performance structural materials based on the plaston concept. Chapters 11 and 12 describe high manganese steels and magnesium alloys, respectively.

This is the first book that comprehensively explains the plaston, an important new concept in structural materials research, from both theoretical and experimental perspectives. Although the study of structural materials has been conducted for many years, there are still new and important problems to be solved. We believe that this book will provide useful information to researchers in both academia and industry who are challenging these issues.

We would like to thank all members of ESISM and the steering committee for their lively discussions and encouragement for the publication of this book. We would also like to express our gratitude to the management office of ESISM at Kyoto University, especially to Prof. Shojiro Ochiai, Prof. Lina Kawaguchi, and Prof. Kiichiro Oishi. This book would not be possible without their devoted efforts. Financial support for the open access publication of this book by MEXT Japan through ESISM of Kyoto University is gratefully acknowledged.

Kyoto, Japan   Isao Tanaka
                                                                         Nobuhiro Tsuji
                                                                         Haruyuki Inui

Contents

Part I
Introduction

Chapter 1
Proposing the Concept of *Plaston* and Strategy to Manage Both High Strength and Large Ductility in Advanced Structural Materials, on the Basis of Unique Mechanical Properties of Bulk Nanostructured Metals

Nobuhiro Tsuji, Shigenobu Ogata, Haruyuki Inui, Isao Tanaka, and Kyosuke Kishida

1.1 Introduction

Higher and higher strengths are nowadays required for structural materials, for reducing the weight of transportation machines like automobiles in order to improve fuel efficiency, realizing huge constructions like ultra-tall towers, and securing human beings and society from incidents and disasters like collisions, earthquakes, and so on. However, the ductility and/or toughness of materials generally decreases with an increase in the strength, as is schematically illustrated in Fig. 1.1. The curve showing the trade-off relationship between strength and ductility/toughness is often called the "banana curve" due to the shape shown in Fig. 1.1 (Demeri 2013). Among three major industrial materials, ceramics are very strong but generally brittle and scarcely show plasticity. Polymers are light and ductile in many cases, but their strength is limited. Only metals can manage both high strength and large plasticity, which is derived from their metallic bonding nature. Even in metallic materials, however, the

N. Tsuji (✉) · S. Ogata · H. Inui · I. Tanaka · K. Kishida
Center for Elements Strategy Initiative for Structural Materials (ESISM), Kyoto University, Sakyo-ku, Kyoto 606-8501, Japan
e-mail: nobuhiro-tsuji@mtl.kyoto-u.ac.jp

N. Tsuji · H. Inui · I. Tanaka · K. Kishida
Department of Materials Science and Engineering, Graduate School of Engineering, Kyoto University, Kyoto 606-8501, Japan

S. Ogata
Department of Mechanical Science and Bioengineering, Graduate School of Engineering Science, Osaka University, Toyonaka 560-8531, Japan

© The Author(s) 2022   3
I. Tanaka et al. (eds.), *The Plaston Concept*,
https://doi.org/10.1007/978-981-16-7715-1_1

Fig. 1.1 Schematic illustration showing the strength-ductility balances in various kinds of materials

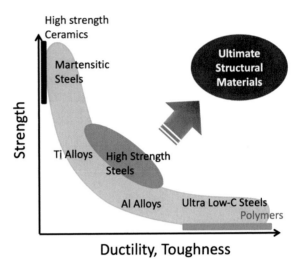

ductility and toughness deteriorate with increasing strength. We have to overcome the trade-off relationship between strength and ductility/toughness for realizing the ultimate structural materials in future, since high strength materials would also be manufactured into designed shapes by metalworking processes and must avoid brittle or early fracture in practical use. Recently a number of articles have claimed to find new metallic materials managing both high strength and good ductility (Zhao et al. 2006; Lu et al. 2009; Copper et al. 2011; Liu et al. 2013, 2018; Wu et al. 2014, 2015; Wei et al. 2014; Kim et al. 2015; Li et al. 2016; He et al. 2017; Lei et al. 2018; Tong et al. 2018; Yang et al. 2018; Sun et al. 2019; Zhang et al. 2019, Ma and Zhu 2017). However, most of the papers have mainly insisted on superior mechanical properties found in particular materials with different (and mostly complicated) microstructures, and the discussions on the reason why those materials could realize high strength and large ductility have stayed in phenomenological manners. We still do not have the guiding principle to manage both high strength and high ductility in advanced structural materials.

The first author of this manuscript has continuously studied bulk nanostructured metals (or ultrafine-grained (UFG) metallic materials, in other words) in the last decades, and has found their unique mechanical properties, such as the unexpected yield-drop phenomena universally found in UFG metals and alloys regardless of their chemical compositions and crystal structures (Tsuji et al. 2002; Tian et al. 2018a, 2020a, b; Saha et al. 2013; Yoshida et al. 2017, 2019; Terada et al. 2008; Zheng et al. 2017, 2019, 2020a; Gao et al. 2014a; Bai et al. 2021), extra Hall–Petch strengthening (Gao et al. 2014a; Kamikawa et al. 2009; Tian et al. 2020), "hardening by annealing and softening by deformation" phenomena (Huang et al. 2006), and so on. Bulk nanostructured metals show very high strength compared to conventionally coarse-grained counterparts, but most of them still have a dilemma of the strength-ductility trade-off (Tsuji et al. 2002, 2008). On the other hand, it has been also found recently that bulk nanostructured metals, in particular, alloys could manage both high strength and ductility (Tsuji et al. 2019; Tian et al. 2014, 2015, 2016, 2018b; Chen et al. 2014; Bai et al. 2018; Gao et al. 2019; Chong et al. 2019; Zhang et al. 2020). In the current manuscript, the authors throw light on the reason why such bulk nanostructured metals can overcome the strength-ductility trade-off, and propose a strategy for managing both high strength and large ductility in advanced structural materials. We also propose a new concept of *"plaston"* for considering the nucleation of different deformation modes (Tsuji et al. 2020, 2021), which can be a key to realizing the strategy.

1.2 Reason of Strength-Ductility Trade-Off, and Mechanical Properties of Typical Bulk Nanostructured Metals

In general, metallic materials used in our society are polycrystals composed of a number of crystal grains having a different crystallographic orientation to each other. It has been well known that refinement of the grain size in the polycrystalline metallic materials improves their mechanical properties such as strength, toughness, etc. However, the minimum mean grain size in bulky metals we can obtain through conventional fabrication processes has been about 10 μm. Since 1990s, great attention has been paid to the ultrafine-grained (UFG) metals, of which average grain sizes are smaller than 1 μm, since various kinds of new processes like severe plastic deformation (SPD) (Altan 2006; Azushima et al. 2008) made it possible to fabricate bulky metallic materials having such ultrafine-grained structures. Equal channel angular extrusion (ECAE), high pressure torsion (HPT), accumulative roll bonding (ARB), etc., are typical SPD processes that can fabricate UFG structures in bulky metals (Altan 2006; Azushima et al. 2008). Figure 1.2 shows the volume fraction of grain boundaries in polycrystalline materials as a function of the mean grain size, assuming that the thickness of the grain boundary regions, where arrangements of atoms are locally distorted from the periodical and well-organized crystalline structures in grain interiors, is 1 nm (Tsuji 2002, 2007). In the polycrystalline materials having average grain sizes over 10 μm, which correspond to the conventional metallic materials human beings have used by now, the volume fraction of grain boundaries is negligibly small. It can be said, therefore, that the conventional metals and alloys scarcely involve grain boundaries, even though they are polycrystalline materials. However, the grain boundary fraction quickly increases with decreasing the grain size below 1 μm. The volume fractions of grain boundaries in the materials with the mean grain sizes of 1 μm, 0.1 μm (100 nm), and 10 nm are 0.2%, 2%, and 20%, respectively. That is, the UFG metals are materials full of grain boundaries. Therefore, it would not be surprising that the UFG or nanocrystalline materials show unique behavior and superior properties that are significantly different from those of the conventional metals having coarse grain sizes over 10 μm. We can call such UFG

Fig. 1.2 Volume fraction of grain boundaries in polycrystalline materials as a function of the mean grain size. It is assumed that the thickness of the grain boundary regions, where atomic arrangements are locally distorted from the periodical and well-organized crystalline structures in grain interiors, is 1 nm

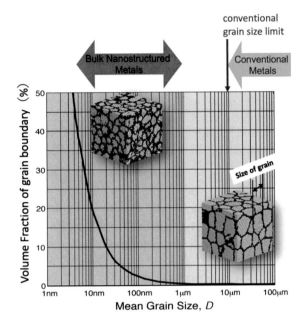

metallic materials having bulky dimensions and average grain sizes of nano-meter scale as "Bulk Nanostructured Metals".

Bulk nanostructured metals usually exhibit very high strengths. Figure 1.3 shows tensile strengths of a commercial purity Al (JIS 1100-Al) and an ultra-low carbon interstitial free (IF) steel (Tsuji 2007). The blue bars in Fig. 1.3 indicate the tensile strengths of starting materials having conventionally coarse average grain sizes ($d \sim$ 20 μm), while red bards represent the tensile strengths of the UFG specimens ($d \sim$

Fig. 1.3 Tensile strength of a commercial purity aluminum (JIS 1100-Al) and IF steel having different mean grain sizes (d). The starting materials had conventionally coarse grain sizes ($d \sim 20$ μm), while the nanostructured specimens of the same materials fabricated by the ARB process showed $d \sim 0.2$ μm

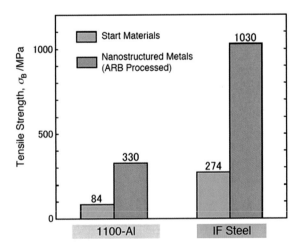

0.2 μm) fabricated by the ARB process. The coarse-grained specimen of 1100-Al showed a tensile strength of 84 MPa, but it increased to 330 MPa when the grain size was reduced down to 0.2 μm. The strength of the UFG Al was four times higher than that of the coarse-grained counterpart and surprisingly higher than the strength of the IF steel with conventional grain size (274 MPa). That is, aluminum can be strengthened as high as steel by the ultra-grain refinement. Such strengthening in nanostructured metals was also found in the IF steel (Fig. 1.3). The typical strength of the coarse-grained starting IF steel specimen (274 MPa) increased over 1 GPa by nano-structuring.

Bulk nanostructured metals show such high strength, but their tensile ductility is limited in most cases. A typical example of the trade-off relationship between strength and tensile ductility in nanostructured metals is shown in Fig. 1.4a, which indicates engineering stress–strain curves of an ultra-low carbon interstitial free (IF) steel (Takechi 1994) having various average grain sizes (*d*) ranging from 0.4 to 33 μm (Gao et al. 2014b). The IF steels have a single-phase *α* matrix with body-centered cubic (BCC) crystal structure. The specimens with various mean grain sizes were fabricated by SPD using the ARB process (Saito et al. 1998, 1999; Tsuji et al. 2003) and subsequent annealing (Gao et al. 2014b). Figure 1.4b shows microstructures of the specimens obtained by electron back-scattering diffraction (EBSD) analysis in

Fig. 1.4 **a** Engineering stress–strain curves of the ultra-low carbon IF steel with different average grain sizes (*d*) ranging from 0.4 μm to 33 μm. The specimens were fabricated by the ARB process followed by annealing (Tsuji et al. 2020a). **b** EBSD-IPF maps showing typical microstructures of the specimens. The colors indicate crystallographic orientations parallel to the rolling direction (RD) (Gao et al. 2014b). Reprinted from Scripta Materialia, vol. 181, Tsuji et al., Strategy for managing both high strength and large ductility in structural materials–sequential nucleation of different deformation modes based on a concept of *plaston*, p. 36, Copyright (2020), with permission from Elsevier

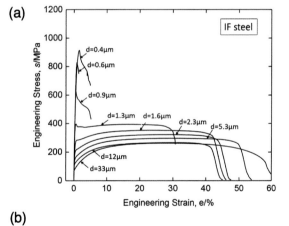

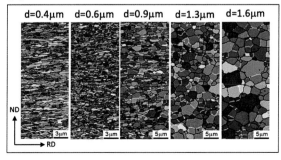

field-emission type scanning electron microscope (FE-SEM). The specimen with d = 0.4 μm was the as-ARB processed specimen, and the EBSD inverse pole figure (IPF) map represented elongated lamellar UFG grains typically found in metals SPD processed in monotonic direction (Tsuji et al. 2019). This is also a kind of deformed microstructure formed by the grain subdivision mechanism (Tsuji et al. 2019), where grains are elongated along the major deformation direction and include deformed substructures with high dislocation densities. By increasing the annealing temperature and time, recovery and grain boundary migration happened to make the dislocation density lower and the grain size larger. The specimens having the mean grain sizes of 1.3 μm or larger showed equiaxed grains free from dislocations, which were equivalent to recrystallized microstructures. The strength, especially the yield strength of the IF steel increased with decreasing grain size (Fig. 1.4a). The relationship between the yield strength and the average grain size followed the well-known Hall–Petch relationship (Hall 1951; Petch 1953). The Hall–Petch relationship is expressed as

$$\sigma_y = \sigma_0 + k d^{-\frac{1}{2}} \qquad (1.1)$$

where σ_y is the yield strength of the material, σ_0 and k are constants depending on materials, and d is the average grain size. Different from the strength, the tensile ductility, especially the uniform elongation of the IF steel suddenly decreased down to a few % when the average grain size became smaller than 1 μm (Fig. 1.4a). This is the typical strength–ductility trade-off observed in most of the bulk nanostructured metals (Tsuji et al. 2002, 2008; Tsuji 2002, 2007). It should be also noted, by the way, that the specimens with grain sizes smaller than 2 μm showed discontinuous yielding characterized by a clear yield-drop phenomenon, even though the IF steels have no interstitial carbon and nitrogen atoms and they normally show continuous yielding (Takechi 1994). In fact, the IF steel specimens with grain sizes larger than 5 μm indicated typical continuous yielding without yield drop (Fig. 1.4a). Fully annealed or recrystallized UFG metals and alloys universally show such a yield-drop phenomenon, regardless of their chemical compositions and crystal structures (Tsuji et al. 2002; Tian et al. 2018a, 2020a; Saha et al. 2013; Yoshida et al. 2017, 2019; Terada et al. 2008; Zheng et al. 2017, 2019, 2020a; Gao et al. 2014a; Bai et al. 2021). As will be argued later, the yield-drop phenomenon found in bulk nanostructured metals is the sign of the activation of the *plaston* (or the activation of new deformation mode).

Another example of typical stress–strain curves of bulk nanostructured metals is represented in Fig. 1.5, which shows engineering (nominal) stress–strain curves of a high-purity (99.99 mass % purity) Al with various average grain sizes (Kamikawa et al. 2009). The specimens were fabricated by ARB and subsequent annealing processes. The as-ARB processed specimen had ultrafine-grained structure with d = 0.88 μm, but still somehow maintained features of deformed microstructures, i.e., elongated grain morphologies and dislocations inside grains (Kamikawa et al. 2009). The specimen showed relatively high yield strength for a high purity metal,

Fig. 1.5 Engineering stress–strain curves of the high-purity (99.99 mass % purity) Al having different mean grain sizes (*d*) ranging from 0.88 to 23 μm. The specimens were fabricated by the ARB process followed by annealing (Kamikawa et al. 2009)

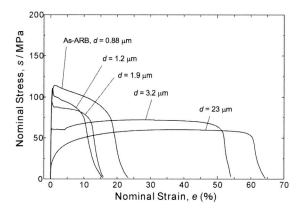

but the uniform elongation was very limited again (1–2%). The total elongation of this specimen was fairly large (>20%), probably because the material had high purity and was soft. Although the specimens with mean grain sizes of 1.2 μm or larger already had nearly equiaxed grain structures with low dislocation density (Kamikawa et al. 2009), the 1.2 and 1.9 μm grain-sized specimens still showed limited uniform elongation. It should be noted again that the specimens having mean grain sizes of 1.2, 1.9, and 3.2 μm showed a sharp yield drop, although the coarse-grained high-purity Al (*d* = 23 μm) exhibited continuous yielding usually observed in FCC metal. Even typical Lüders deformation was found in the 3.2 μm grain-sized specimen, which corresponded to the nearly flat part on the stress–strain curve. After the Lüders deformation, the specimen showed a slight strain hardening, so that tensile ductility (uniform elongation) was significantly recovered in the 3.2 μm grain-sized specimen although the strength already decreased very much. These are basically the same characteristics as those of the UFG IF steel with BCC structure shown in Fig. 1.4. The outlines of the stress–strain curves shown in Figs. 1.4a and 1.5 show banana-like shapes that correspond to the strength–ductility trade-off curve shown in Fig. 1.1.

The sudden drop of uniform elongation and tensile ductility in bulk nanostructured metals is simply explained by the plastic instability (Tsuji et al. 2002; Tsuji 2007; Wang et al. 2002; Morris 2008). In tensile testing of materials, necking of the gage part may happen. Since the cross-sectional area of the necked region becomes smaller than the un-necked region, the tensile stress at the necked region becomes higher. On the other hand, the necked region becomes harder than the un-necked region, since the necked region is much more plastically deformed and strain-hardened than the un-necked region. Consequently, whether the necking progresses furthermore or not is determined by the balance between the increased tensile stress and increased strength of the necked region, which is known as the plastic instability condition. Considère criterion shown below is a well-known plastic instability condition for strain-rate insensitive materials (Wagoner and Chenot 1997).

$$\left(\frac{d\sigma}{d\epsilon}\right) \leq \sigma \tag{1.2}$$

Here, σ is the true flow stress and ε is the true strain, so that $(d\sigma/d\varepsilon)$ corresponds to the strain-hardening rate. The plastic instability condition indicates the propagation of necking, that is, the uniform elongation in tensile testing. The Eq. (1.2) tells that the strain-hardening rate $(d\sigma/d\varepsilon)$ plays a critical role in the uniform elongation of materials. Grain refinement of metallic materials primarily increases the yield strength and flow stress (σ) according to the Hall–Petch relationship (Hall 1951; Petch 1953) shown in the Eq. (1.1), but the strain-hardening ability $(d\sigma/d\varepsilon)$ of the materials is not enhanced because the structure at grain interiors does not change by the grain refinement. As a result, plastic instability would happen at an earlier stage of tensile deformation in materials with finer grain sizes. Such a situation is schematically illustrated in Fig. 1.6. The figure indicates the change of the plastic instability points when the yield strength increases keeping a constant strain-hardening rate, which clearly illustrates that the plastic instability occurs at earlier stages of tensile tests in the material having the finer grain sizes.

The plastic instability condition in UFG materials is examined for the IF steel with various mean grain sizes (d) of which tensile properties are shown in Fig. 1.4a. Figure 1.7 shows true stress–strain curves of the IF steel and strain-hardening curves of the specimens having grain sizes of $d = 33$ μm (blue) and $d = 0.4$ μm (red). The stress–strain curves are drawn in solid lines till the points of uniform elongation which is determined on the corresponding engineering stress–strain curves, and then drawn in broken lines after the uniform elongation. Generally, the strain-hardening rate of metallic materials monotonically decreases with increasing the plastic strain and meets with the true stress–strain curve. The meeting point corresponds to the

Fig. 1.6 Schematic illustration showing the change of the plastic instability point (i.e., the point of uniform elongation) by grain refinement. It is assumed that the yield strength increases by the grain refinement while the strain-hardening rate ($d\sigma/d\varepsilon$) does not change

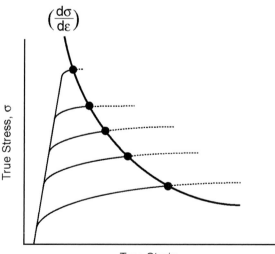

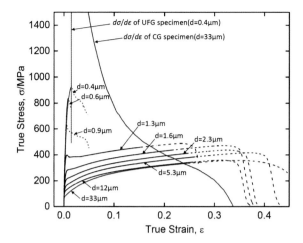

Fig. 1.7 True stress–strain curves of the IF steel specimens having different grain sizes and and strain-hardening rate ($d\sigma/d\varepsilon$) curves of the specimens with the average grain sizes of 33 and 0.4 μm. The stress–strain curves are drawn in solid lines up to the uniform elongation points determined from the engineering stress–strain curves, and then drawn in broken lines in post uniform elongation regions after. Curves for the representatively coarse grained ($d = 33$ μm) and ultrafine grained ($d = 0.4$ μm) specimen are expressed in blue and red, respectively (Tsuji et al. 2020a). Reprinted from Scripta Materialia, vol. 181, Tsuji et al., Strategy for managing both high strength and large ductility in structural materials–sequential nucleation of different deformation modes based on a concept of *plaston*, p. 36, Copyright (2020), with permission from Elsevier

plastic instability. The blue curves for the coarse-grained ($d = 33$ μm) IF steel typically show such a behavior. The intersection of two curves (the plastic instability point) coincides well with the uniform elongation point where the curve changes from the solid line to the broken line. For the UFG specimen with $d = 0.4$ μm, the red strain-hardening rate curve quickly falls and meets with the flow stress curve near the maximum stress, which also indicates that the plastic instability point coincides with the uniform elongation point. The results confirm that the uniform elongation of these specimens can be well explained by the plastic instability. One of the reasons for the quick decrease of the strain-hardening rate is localized deformation appearing with the yield drop. However, the strain-hardening rate of the 0.4 μm specimen does not recover again, while that of the 1.3 μm specimen increases again after the yield drop and Lüders deformation. Anyway, Fig. 1.7 obviously confirmed the early plastic instability in fine-grained materials, as was schematically interpreted in Fig. 1.6. It seems that the limited uniform elongation (i.e., limited tensile ductility) due to the early plastic instability is indispensable in bulk nanostructured metals having UFG structures.

1.3 Bulk Nanostructured Metals Exhibiting Both High Strength and Large Ductility

Although many bulk nanostructured metals exhibit the early plastic instability as shown in the former Sect. 1.2, it has been found recently that some kinds of alloys with UFG microstructures can manage both high strength and large tensile ductility (Tsuji et al. 2019; Tian et al. 2014, 2015, 2016, 2018b; Chen et al. 2014; Bai et al. 2018, 2021; Gao et al. 2019; Chong et al. 2019; Zhang et al. 2020). One example was found in a Mg–6.2Zn–0.5Zr–0.2Ca (mass %) alloy (ZKX600) (Zheng et al. 2017, 2019). Figure 1.8 shows EBSD IPF and grain boundary (GB) maps of the ZKX600 specimens severely deformed by the high-pressure torsion (HPT) process and then annealed. Fully recrystallized microstructures with various average grain sizes ranging from 0.77 μm to 23.3 μm were obtained. Figure 1.9 shows engineering (nominal) stress–strain curves of the Mg–Zn–Zr–Ca specimens having various mean grain sizes ranging from 0.1 to 23.3 μm. The 0.1 μm specimen was just HPT processed, which showed very high yield strength but limited tensile elongation. These are typical mechanical properties of as-SPD processed materials, which could be also found in Figs. 1.4a and 1.5. Other specimens in Fig. 1.9 were all fully recrystallized, even in case of the grain size of 0.77 μm, as shown in Fig. 1.8 (Zheng et al. 2017, 2019). The strength of the alloy continuously increased with decreasing the grain size. It was interesting that the tensile elongation rather increased with decreasing

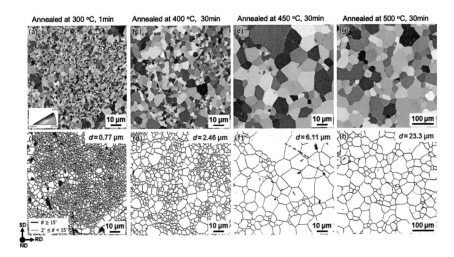

Fig. 1.8 EBSD IPF maps (**a, c, e, g**) and grain boundary (GB) maps (**b, d, f, h**) of the ZKX600 Mg alloy heavily deformed by high pressure torsion (HPT) and then annealed under different conditions. **a, b** Annealed at 300 °C for 60 s; **c, d** 400 °C for 1.8 ks; **e, f** 450 °C for 1.8 ks; **g, h** 500 °C for 1.8 ks. Colors in the IPF maps indicate crystallographic orientations parallel to the normal direction (ND) of the disc specimens. In the GB maps, high angle boundaries with misorientations (θ) larger than 15° and low angle boundaries with $2° \leq \theta < 15°$ are drawn in blue and green lines, respectively (Zheng et al. 2017, 2019)

Fig. 1.9 Engineering stress–strain curves of the Mg–Zn–Zr–Ca alloy (ZKX600) with various mean grain sizes (*d*) ranging from 0.1 to 23.3 μm, fabricated by HPT and subsequent annealing (Zheng et al. 2017, 2019)

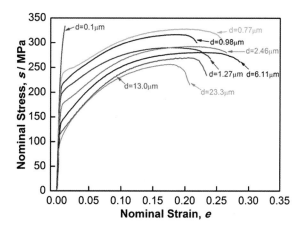

the average grain size from 23.3 to 6.11 μm. When the grain size decreased furthermore, the elongation somehow decreased but still maintained fairly large values over 20%. That is, both high strength and large tensile ductility were managed in the UFG Mg–Zn–Zr–Ca alloy. The 23.3 μm grain-sized specimen showed typical continuous yielding, while the yielding behavior changed to discontinuous type with the grain refinement. Although clear yield drop was not observed in Fig. 1.9, the local strain analysis by the digital image correlation (DIC) technique confirmed that the 0.77 μm grain-sized specimen showed localized deformation similar to Lüders banding around yielding (Zheng et al. 2019).

For comparing mechanical properties, two specimens with coarse and ultrafine grain sizes (*d* = 23.3 μm and 0.77 μm, respectively) were selected. The grain refinement from 23.3 μm to 0.77 μm significantly increased the yield strength of the ZKX600 Mg alloy from 90 to 235 MPa, respectively (Zheng et al. 2017). As mentioned above, the coarse-grained specimen showed continuous yielding typically observed in Mg alloys with conventionally coarse grain sizes, while the UFG specimen showed a discontinuous shape. Different from the IF steel (Fig. 1.4) and pure Al (Fig. 1.5), the UFG Mg alloy showed good strain hardening after yielding, so that it exhibited higher tensile strength (328 MPa) as well as larger uniform and total elongations (20.5% and 26.1%, respectively) than the coarse-grained specimen (showing the tensile strength, uniform elongation and total elongation of 256 MPa, 17.6%, and 20.3%, respectively).

True stress–strain curves and strain-hardening rate curves of the specimens with *d* = 23.3 μm and *d* = 0.77 μm are shown in Fig. 1.10 (Zheng et al. 2019; Tsuji et al. 2020a), in order to examine the plastic instability conditions. The curves for the coarse-grained specimen were drawn in blue, while those for the ultrafine-grained specimen were drawn in red. The strain-hardening rate ($d\sigma/d\varepsilon$) of the UFG specimen (*d* = 0.77 μm; red curve) quickly decreased at the beginning of tensile deformation, which corresponded to the discontinuous yielding of this specimen. However, the

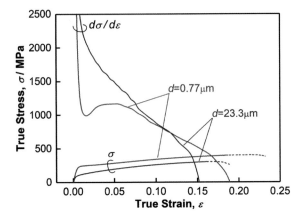

Fig. 1.10 True stress–strain curves and strain-hardening rate ($d\sigma/d\varepsilon$) curves of the ZXK600 Mg alloy specimens with average grain sizes of 23.3 μm (blue) and 0.77 μm (red) (Zheng et al. 2019; Tsuji et al. 2020a). Reprinted from Scripta Materialia, vol. 181, Tsuji et al., Strategy for managing both high strength and large ductility in structural materials–sequential nucleation of different deformation modes based on a concept of *plaston*, Copyright (2020), with permission from Elsevier

strain-hardening rate recovered, and showed high values comparable to the strain-hardening rate of the coarse-grained specimen ($d = 23.3$ μm; blue curve). At later stages of the tensile deformation, the strain-hardening rate of the UFG specimen maintained higher values than that of the coarse-grained specimen, which explained the postponement of plastic instability and the larger uniform elongation of the UFG specimen than that of the coarse-grained specimen.

Why did the UFG specimen of the Mg alloy exhibit both high strength and large tensile ductility? In order to clarify this, deformation mechanisms in the Mg alloy were investigated. It is well known that basal slip having Burgers vector (***b***) parallel to an <***a***> axis in hexagonal close-packed (HCP) crystal structure is the easiest deformation mode in HCP Mg and its alloys (Yoshinaga and Horiuchi 1964). However, basal slips can realize only two-dimensional deformation in each grain, since there is only one basal plane in HCP crystal. As a result, deformation twinning having deformation components along the ***c***-axis operates in Mg and its alloys. Area fractions of deformation twins in the current specimens tensile-deformed to different strains were measured by crystallographic analysis using EBSD. Figure 1.11 shows the measured area fraction of the deformation twins in the specimens having average grain sizes of 23.3 and 0.98 μm (Zheng et al. 2019). In the coarse-grained specimen ($d = 23.3$ μm), the fraction of deformation twins increased with increasing the tensile strain, which is a typical behavior in Mg alloys. On the other hand, deformation twinning was significantly suppressed in the UFG specimen ($d = 0.98$ μm). Even after 13% tensile strain, the area fraction of deformation twins was only ~2%. In such a case, how was the three-dimensional plastic deformation without incompatibility between neighboring grains realized in the UFG specimens? For clarifying this

Fig. 1.11 Area fractions of deformation twins in the ZKX600 Mg alloy specimens with average grain sizes of 23.3 and 0.98 μm (Zheng et al. 2019)

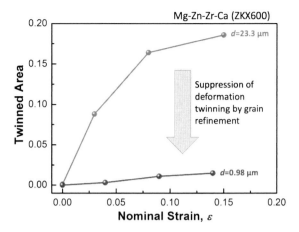

point, dislocation substructures were observed by transmission electron microscopy (TEM). Figure 1.12 shows TEM images of the UFG specimen ($d = 0.77$ μm) of ZKX600 Mg alloy tensile-deformed to 9.5% strain (Zheng et al. 2019; Tsuji et al. 2020a). A number of dislocations are observed in the ultrafine grain located in the bright-field image observed from near [01–10] zone axis (Fig. 1.12a). Figure 1.12b shows a dark-field image of the region surrounded by the red-broken rectangle in (**a**). Since $g = 0002$ was used for the dark-field observation, contrasts of dislocations belonging to basal slip systems with only <*a*> component of Burgers vector (**b**) must disappear, according to the $g \cdot b = 0$ criterion. However, many dislocations are still observed clearly in Fig. 1.12b, which indicates there are many dislocations with <*c*>

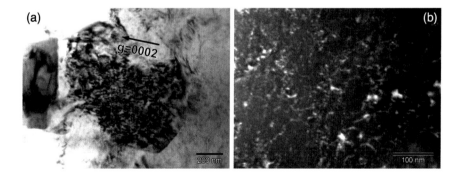

Fig. 1.12 TEM images of the UFG specimen ($d = 0.77$ μm) of ZKX600 Mg alloy tensile-deformed to 9.5% engineering strain. **a** Bright-field image observed from near [01–10] zone axis under two-beam condition. **b** Dark-field image of the region surrounded by red-broken rectangle in (**a**), using $g = 0002$ (Zheng et al. 2019; Tsuji et al. 2020a). Reprinted from Scripta Materialia, vol. 181, Tsuji et al., Strategy for managing both high strength and large ductility in structural materials–sequential nucleation of different deformation modes based on a concept of *plaston*, Copyright (2020), with permission from Elsevier

Fig. 1.13 Engineering stress–strain curves of the 31Mn–3Al–3Si (mass%) steel with various mean grain sizes (*d*) ranging from 0.79 to 85.6 μm, fabricated by conventional heavy cold-rolling and subsequent annealing (Kitamura 2017; Bai et al. 2016, 2021)

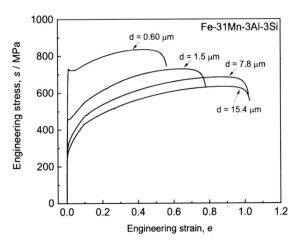

components of Burgers vector within the ultrafine grain. They are presumably dislocations of pyramidal slip systems with <*c* + *a*> Burgers vector. It can be concluded that deformation twinning was greatly suppressed in the UFG Mg alloy, and instead of deformation twins, <*c* + *a*> dislocations were activated to realize three-dimensional plastic deformation for satisfying deformation compatibility in the polycrystalline material. The easy basal slips were also activated in all specimens regardless of the grain size. Dislocations belonging to such different slip systems and having different Burgers vectors would interact with each other, which would then inhibit dynamic recovery by cross-slips of screw dislocations, leading to a significant increase of dislocation density and strain-hardening rate, as was shown in Fig. 1.10. The unexpected activation of unusual <*c* +*a*> dislocation is the reason for the regeneration of strain hardening and overcoming the trade-off relationship between strength and ductility in the bulk nanostructured Mg alloy.

Next, another type of example for the UFG materials overcoming the strength–ductility trade-off is introduced. Figure 1.13 shows engineering stress–strain curves of a 31Mn–3Al–3Si (mass %) steel with different mean grain sizes ranging from 0.60 to 15.4 μm (Kitamura 2017; Bai et al. 2016, 2021). The 31Mn–3Al–3Si steel has a stable austenite (face-centered cubic (FCC) crystal structure) single-phase microstructure at room temperature. High-Mn steels are known to manage both high strength and good tensile ductility. It has been considered that their good mechanical properties are attributed to deformation twinning frequently happening during plastic deformation, so that they are called twinning-induced plasticity (TWIP) steels (Grässel et al. 2000; De Cooman et al. 2018). The 31Mn–3Al–3Si steel is one of the typical high-Mn TWIP steels. The specimens having various average grain sizes (*d*) ranging from 0.60 to 15.4 μm were fabricated by conventional heavy cold rolling and subsequent annealing, and all the specimens showed fully recrystallized microstructures. As is shown in Fig. 1.13, the strength of the alloy increased with decreasing grain size. Tensile ductility showed nearly the same large values (~100%) from *d* = 15.4 μm to *d* = 7.8 μm, and then gradually decreased with decreasing the grain size. However, even the ultrafine-grained specimen with $d = 0.60$ μm showed large tensile elongation (uniform elongation) over 40%. The 31Mn–3Al–3Si steel showed single-phase austenite structure, so that coarse-grained specimens exhibited continuous yielding typically found in FCC metals and alloys. On the other hand, discontinuous yielding with clear yield drop was found in the UFG specimens with mean grain sizes smaller than 1.5 μm. Stress–strain curve of the 0.60 μm specimen showed a flat part corresponding to Lüders deformation, although the Lüders strain was small (~4%).

True stress–strain curves and corresponding strain-hardening rate curves of the Fe–31Mn–3Al–3Si specimens having coarse grain size ($d = 15.4$ μm; blue) and ultrafine grain size ($d = 0.60$ μm; red) are shown in Fig. 1.14 (Tsuji et al. 2020a). Similar to the IF steel (Fig. 1.7) and the ZKX600 Mg alloy (Fig. 1.10), the strength of the austenitic steel significantly increased by the grain refinement. The strain-hardening rate ($d\sigma/d\varepsilon$) of the UFG specimen with $d = 0.60$ μm quickly dropped, corresponding to the discontinuous yielding with the yield drop, but promptly recovered and maintained rather higher values than that of the coarse-grained specimen. Consequently, the UFG specimen showed high tensile strength (836 MPa) and large uniform elongation (43.3%).

In order to understand the deformation mechanism, deformation microstructures of the Fe–31Mn–3Al–3Si specimens tensile-deformed were carefully observed. First, the specimens having average grain sizes of 15.4 μm (coarse grained), 4.5 μm (fine-grained), and 0.79 μm (ultrafine grained) were deformed to different strains by

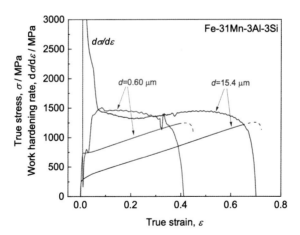

Fig. 1.14 True stress–strain curves and corresponding strain-hardening rate curves of the 31Mn–3Al–3Si austenitic steel specimens having coarse grain size ($d = 15.4$ μm; blue) and ultrafine grain size ($d = 0.60$ μm; red) (Tsuji et al. 2020a). Reprinted from Scripta Materialia, vol. 181, Tsuji et al., Strategy for managing both high strength and large ductility in structural materials–sequential nucleation of different deformation modes based on a concept of *plaston*, Copyright (2020), with permission from Elsevier

Fig. 1.15 Number of deformation twins per unit area obtained from ECC images of the tensile tensile-deformed specimens having different mean grain sizes of 15.4 μm, 4.5 μm, and 0.79 μm. Plotted as a function of tensile engineering strain (Bai et al. 2021)

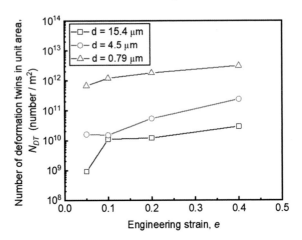

tensile tests, and then deformation substructures were observed by electron channeling contrast imaging (ECCI) in SEM. From ECCI data, area fraction and number density of deformation twins were measured. Figure 1.15 summarizes changes of the number densities [number/m^2] of deformation twins with a progress of tensile deformation. It has been experimentally shown and believed that grain refinement of matrix suppresses deformation twinning in FCC metals and alloys (Surya et al. 1999). The suppression of deformation twinning by grain refinement has been explained by the decrease of the chance of dislocation reactions that produce deformation twins in FCC crystals (Venables 1961; Cohen and Weertman 1963; Miura et al. 1968; Mahajan and Chin 1975). However, Fig. 1.15 clearly shows that the number of deformation twins per unit area increases by the grain refinement. Especially, the UFG specimen with $d = 0.79$ μm exhibited much larger number of twins than the conventionally coarse-grained specimen with $d = 15.4$ μm.

An ECC image of the UFG specimen with $d = 0.79$ μm tensile-deformed to 1.6% is shown in Fig. 1.16 (Tsuji et al. 2020a; Kitamura 2017; Bai et al. 2021). At this early stage of tensile deformation just after the yield drop, thin deformation twins nucleated from grain boundaries of ultrafine grains. The deformation microstructures of the 31Mn–3Al–3Si specimens with different mean grain sizes were also observed by TEM carefully (Hung et al. 2021a, b, c). TEM images shown in Fig. 1.17 indicate a transition of deformation substructures from in-grain slip dislocations in a coarser grain to deformation twins nucleated from grain boundaries in an ultrafine grain (Hung et al. 2021a). It should be noted that the deformation twins are quite thin and have thicknesses of a few nm ~ 10 nm. As a result, the area fraction of deformation twins in the UFG specimen was rather smaller than that of the coarse-grained specimen. Since twin boundaries are high-angle grain boundaries (mostly $\Sigma 3$ boundaries), they act as strong obstacles for dislocation slips. The formation of a number of thin twins subdivides the matrix very much and would decrease the slip length of dislocations, leading to more dislocations accumulated. This would be the reason why the strain hardening of the UFG 31Mn–3Al–3Si specimens was

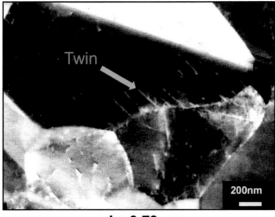

d = 0.79 μm

Fig. 1.16 SEM-ECCI picture of the Fe–31Mn–3Al–3Si specimen having an UFG grain size ($d =$ 0.79 μm) after 1.6% tensile deformation. Nucleation of thin deformation twins from grain boundaries is observed (Tsuji et al. 2020a; Kitamura 2017; Bai et al. 2021). Reprinted from Scripta Materialia, vol. 181, Tsuji et al., Strategy for managing both high strength and large ductility in structural materials–sequential nucleation of different deformation modes based on a concept of *plaston*, Copyright (2020), with permission from Elsevier

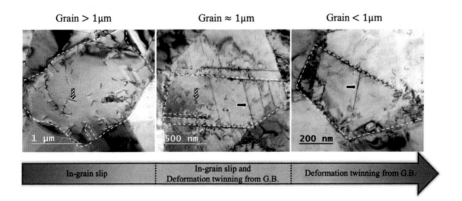

Fig. 1.17 TEM images showing the transition of deformation microstructures from in-grain slip dislocations in a coarse grain to deformation twins nucleated from grain boundaries in an ultrafine grain (Hung et al. 2021a)

enhanced and good mechanical properties managing both high strength and large tensile ductility (uniform elongation) was realized. It should be emphasized again that deformation twinning was exceptionally enhanced and the twins nucleated from grain boundaries (not by the dislocation interactions in grains) in the UFG specimens of the 31Mn–3Al–3Si steel.

The last example introduced as a bulk nanostructured metal managing both high strength and large tensile ductility is a 24Ni–0.3C (mass %) steel. Although the Fe–24Ni–0.3C alloy also shows FCC single phase at room temperature, the austenite is a metastable phase and can transform to α' martensite with body-centered cubic or tetragonal (BCC or BCT) crystal structure by deformation. Fully recrystallized specimens of the 24Ni–0.3C steel composed of single-phase FCC austenite having different average grain sizes were fabricated by SPD using high-pressure torsion (HPT) followed by heat treatments under different conditions (Chen et al. 2014, 2015). Figure 1.18 shows engineering (nominal) stress–strain curves of the coarse-grained ($d = 35\ \mu m$), fine-grained ($d = 1.1\ \mu m$), and UFG ($d = 0.5\ \mu m$) specimens of the 24Ni–0.3C steel. In this metastable austenitic steel, even the coarse-grained specimen ($d = 35\ \mu m$) showed a high strain-hardening rate, leading to high strength and outstanding tensile ductility over 240% elongation. The good mechanical property is attributed to the deformation-induced martensitic transformation, which is called transformation-induced plasticity (TRIP) (Zackay et al. 1967). It is known that the

Fig. 1.18 a Engineering (nominal) stress–strain curves of the 24Ni–0.3C specimens having three different grain sizes, 35 μm, 1.1 μm, and 0.5 μm (Chen 2015). The 35 μm, 1.1 μm, and 0.5 μm grain-sized specimens were tensile-tested at 50 °C, 30 °C, and 20 °C, respectively, and the test temperatures corresponded to the temperature at which the specimens showed the largest elongation. **b** Fractions of martensite in the specimens with three different mean grain sizes tensile tested to different strains (Chen 2015)

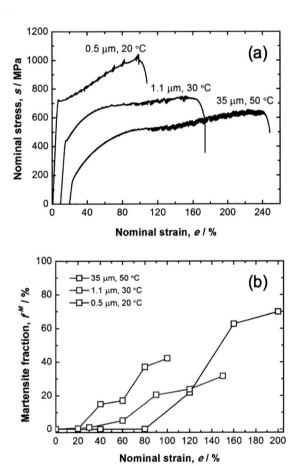

tensile elongation of TRIP steels greatly depends on the deformation temperature since the stability of austenite against martensitic transformation changes depending on temperature, and the maximum elongation is obtained at a certain deformation temperature. The stress–strain curves shown in Fig. 1.18 are those showing the maximum elongations in the different grain-sized specimens. The tensile tests were carried out at different temperatures indicated, because grain refinement stabilizes austenite against martensitic transformation (Umemoto and Owen 1974). Actually, the martensitic transformation starting temperature (*Ms*) upon cooling of austenite decreased from −26 to −66 °C with refining the austenite grain size from 35 to 0.5 μm (Chen et al. 2014; Chen 2015). The strength of materials greatly increased with decreasing the grain size. The coarse-grained specimen ($d = 35$ μm) showed typical continuous yielding, while the fine-grained ($d = 1.1$ μm) and ultrafine-grained ($d = 0.5$ μm) specimens exhibited discontinuous yielding with clear yield-drops as was observed in the UFG IF steel (Fig. 1.4a), pure Al (Fig. 1.5), ZKX600 Mg alloy (Fig. 1.9), and 31Mn–3Al–3Si austenitic steel (Fig. 1.13). Although the tensile ductility decreased with decreasing the grain size, the 0.5 μm grain-sized specimen maintained fairly large tensile elongation over 100% still. Serrations were obviously observed on the stress–strain curves of all the specimens. Such serrations have been considered to correspond to the deformation-induced martensitic transformation during the deformation, although the exact mechanism of serration behavior in TRIP steels is still unclear (Hwang et al. 2021). Fractions of transformed martensite in these specimens tensile tested to different strains were measured by EBSD and plotted in Fig. 1.18b (Chen 2015). The figure clearly indicates that the deformation-induced martensitic transformation was rather enhanced in the UFG specimen, despite that the austenite was stabilized by the grain refinement of austenite, as was described above. The increase of martensite fraction shown in Fig. 1.18b coincided well with the appearance of serrations on the stress–strain curves shown in Fig. 1.18a.

True stress–strain curves and corresponding strain-hardening rate curves of the Fe–24Ni–0.3C steel having average grain sizes of 35 μm (blue) and 0.5 μm (red) are shown in Fig. 1.19 (Tsuji et al. 2020). The specimens with fully recrystallized austenitic microstructures with different mean grain sizes were fabricated by heavy deformation and subsequent heat treatments (Chen et al. 2014), and the tensile tests were conducted at room temperature. The strain-hardening rate of the coarse-grained specimen showed a typical S-shape characteristic in TRIP steels. That is, strain-hardening rate was regenerated by the deformation-induced martensitic transformation. Martensite in carbon-containing steels is much harder than austenite. The strain-hardening rate of the UFG specimen quickly dropped due to the discontinuous yielding again, but gradually recovered to show an S-curve. Although the increase of the strain-hardening rate in the UFG specimen was moderate, compared to that in the coarse-grained specimen, a high strain-hardening rate was maintained up to the later stage of deformation. As a result, the plastic instability point of the UFG specimen was rather postponed than that of the coarse-grained specimen. It has been confirmed that deformation-induced martensite in the UFG specimen nucleated from grain boundaries of austenite matrix (Tsuji et al. 2020a; Mao 2021).

Fig. 1.19 True stress–strain curves and corresponding strain-hardening rate curves of the Fe–24Ni–0.3C steel having average grain sizes of 35 μm (blue) and 0.5 μm (red) (Tsuji et al. 2020)

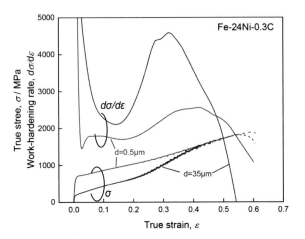

In conclusion, deformation-induced martensitic transformation was unexpectedly activated in the UFG 24Ni–0.3C metastable austenitic steel, which caused significant regeneration of strain-hardening rate, leading to managing both high strength and large ductility. The deformation-induced martensite nucleated from grain boundaries of austenite. The result is quite similar to the unexpected activation of <*c* + *a*> dislocations in the ZKX600 Mg alloy (Fig. 1.12), and unusual enhancement of deformation twinning in ultrafine-grained 31Mn–3Al–3Si austenitic steel (Figs. 1.15, 1.6 and 1.17) shown above, suggesting that there is a key phenomenon commonly happening in bulk nanostructured metals (or UFG metallic materials) for overcoming the strength-ductility trade-off.

1.4 Proposing the Concept of *Plaston* and a Strategy to Overcome Strength-Ductility Trade-Off

In the former Sect. 1.3, mechanical properties of several kinds of bulk nanostructured metals were introduced. Interestingly, some kinds of bulk nanostructured metals managed both high strength and large ductility, which suggested a possibility to overcome the strength-ductility trade-off shown in Fig. 1.1. In such bulk nanostructured metals, different deformation modes were unexpectedly activated, i.e., unusual <*c* + *a*> dislocations in the ZKX600 Mg alloy, deformation twins having nano-thicknesses in the 31Mn–3Al–3Si steel, and deformation-induced martensite in the 24Ni–0.3C metastable austenitic steel. They enhanced or regenerated the strain-hardening rate of the materials, leading to high strength and large ductility. Then, the question is, why were such unexpected deformation modes activated?

The yield-drop phenomena or discontinuous yielding universally happened in bulk nanostructured metals would be the key phenomenon to understanding the activation of unexpected deformation modes. As shown in Figs. 1.4a, 1.5, 1.9, 1.13, and 1.18a, the IF steel, pure Al, ZKX600 Mg alloy, 31Mn–3Al–3Si steel, and 24Ni–0.3C steel all showed discontinuous yielding characterized by clear yield drop in most cases, even though the coarse-grained counterparts of the same alloys exhibited continuous yielding. Here, it should be noted that the UFG specimens that represented the discontinuous yielding had recrystallized microstructures. We consider that this was because the normal deformation mode (i.e., normal dislocation slips) became difficult to operate within fully recrystallized ultrafine grains having limited volumes (Bai et al. 2021; Tsuji et al. 2017). Similar phenomena have been reported after micropillar experiments in the last decade (Uchic et al. 2004, 2009; Greer et al. 2005; Kraft et al. 2010). When the size of the micro-pillar single crystals of various kinds of materials fabricated by focused ion beam (FIB) processing decreased below a few micro-meters, their strengths greatly increased and the "strain-burst" was frequently observed on the load–displacement curves under load-controlled deformation (Uchic et al. 2004, 2009; Greer et al. 2005; Kraft et al. 2010; Fujimura et al. 2011; Inoue et al. 2013; Chen et al. 2016; Okamoto et al. 2016; Zhang et al. 2017; Higashi et al. 2018), which also corresponded to the "pop-in" phenomenon in nano-indentation tests (Li et al. 2021). These have been explained by the so-called dislocation source hardening (Greer et al. 2005; Uchic et al. 2009; Parthasarathy et al. 2007; Lee and Nix 2012). Metallic materials maintain a relatively large number of dislocations even after annealing, so that coarse crystals generally involve easy dislocation sources (like Frank-Read source (Anderson et al. 2017) as well as pre-existing mobile dislocations. Therefore, coarse crystals need not newly nucleate dislocations to initiate and continue plastic deformation. When the crystal size decreases very much, the fine crystals might not include any dislocation sources stochastically, leading to high yield strength sometimes approaching the ideal strength of the crystal (Uchic et al. 2009; Bei et al. 2008; Zhu et al. 2008, 2009; Jennings et al. 2013; Sudharshan Phani et al. 2013; Ogata et al. 2002, 2004; Zhu and Li 2010). It would be reasonable to consider that a similar thing happens in annealed polycrystalline grains with sub-micrometer grain sizes. We think that the yield-drop phenomenon and discontinuous yielding that have been universally observed in fully recrystallized bulk nanostructured metals regardless of the kind of materials (Tsuji et al. 2002; Tian et al. 2018a, 2020a; Saha et al. 2013; Yoshida et al. 2017, 2019; Terada et al. 2008; Zheng et al. 2017, 2019, 2020a; Gao et al. 2014a) reflects such a situation, since it is well known that the yield-drop phenomenon generally happens when free (mobile) dislocations are deficient in crystals (Hall 1970).

The dislocation density in recrystallized metallic materials is known to be about 10^{12} m^{-2}–10^{13} m^{-2}. Such a dislocation density does not probably depend on the grain size. However, as is summarized in Table 1.1, the number of dislocations existing in each grain decreases with decreasing the average grain size naturally. For the calculation, it was simply assumed that grains had cubic shapes and the dislocation density (ρ) was 10^{12} m^{-2} or 10^{13} m^{-2}. It is clearly seen from Table 1.1 that the number of dislocations per grain is significantly limited in ultrafine grain sizes smaller than 1 μm, while each grain has more than 100 dislocations in conventionally coarse-grained materials even in the annealed state. In such coarse-grained polycrystals, loading would induce slips of pre-existing dislocations in each

Table 1.1 The number of dislocations per grain in materials having different average grain sizes (d). It is assumed that the dislocation density (ρ) in the polycrystalline materials is 10^{12} m^{-2} or 10^{13} m^{-2}

Average grain size, d/μm		100	50	10	5	1	0.5	0.1
Number of dislocations per grain	$\rho = 10^{12}$ m^{-2}	10,000	2,500	100	25	1	0.25	0.01
	$\rho = 10^{13}$ m^{-2}	100,000	25,000	1,000	250	10	2.5	0.1

grain. Dislocations can be multiplied by pre-existing Frank-Read sources as well as dislocation sources formed through the enhanced slips (like double cross-slip) of dislocations. These lead to the continuous yielding in coarse-grained polycrystalline metals. On the other hand, when the average grain size decreases to 1 μm, the number of dislocations per grain would be only one or ten under the dislocation density of 10^{12} m^{-2} or 10^{13} m^{-2}, respectively. Slip of such small number of pre-existing dislocations cannot induce large-scale plastic deformation continuously. Furthermore, there would be less chance of dislocation interactions in each tiny grain for creating dislocation sources that can multiply dislocations and realize plastic deformation continuously. As a result, it is difficult to start macroscopic yielding in such ultrafine grains having recrystallized microstructures. For initiating plastic deformation in such nanostructured metals, any carriers of plastic deformation have to be nucleated. When the applied stress (or a local stress) reaches a critical value, such deformation carriers are nucleated and plastic deformation starts. In case of micro pillars and nano-indentation, nucleation of deformation modes initiates most likely from the surfaces (Okamoto et al. 2013, 2014; Nakatsuka et al. 1760; Kishida et al. 2018), and unusual deformation modes might be activated. In fact, Kishida et al. (2020) have recently found a change of deformation mode from deformation twinning into $\{10\bar{1}1\}$ pyramidal $<c + a>$ dislocations in [0001]-oriented micro pillars of pure Ti due to the small specimen size. Deformation modes are possibly activated from grain boundaries in the bulk nanostructured metals of which densities of grain boundaries are high, which agreed with the experimentally observed nucleation of deformation twins in the UFG 31Mn–3Al–3Si steel (Figs. 1.16 and 1.17) and deformation-induced martensite in the UFG 24Ni–0.3C steel from grain boundaries (Tsuji et al. 2020a; Mao 2021). The high stress condition that resulted from the suppression of normal dislocation slips due to the deficiency of pre-existing dislocations and dislocation sources in each recrystallized ultrafine grain is also a key factor for activating the unexpected deformation modes. Similar things must happen in the UFG IF steel and pure Al as well. However, only the same deformation mode as that in the coarse-grained counterparts (i.e., normal a/2 <111> dislocations and a/2 <110> {111} dislocations in BCC and FCC crystals, respectively) operates, so that the strain-hardening rate can not be regenerated in the IF steel and pure Al. Consequently, it can be concluded that activating different deformation modes is important for enhancing or regenerating the strain-hardening rate of the material, in order to realize both high strength and large tensile ductility.

After the consideration mentioned above, a general strategy to manage both high strength and large tensile ductility can be proposed, as schematically represented in Fig. 1.20. The operation of a normal deformation mode, such as conventional dislocation slips, promotes strain hardening. However, in general, the strain-hardening rate monotonically decreases with the progress of plastic deformation, as was shown in Figs. 1.6 and 1.7. If a different deformation mode (deformation mode-2 in Fig. 1.20) is activated, the strain-hardening ability of the material could be regenerated probably owing to interactions between different deformation modes (in other words, interactions between resultant lattice defects of different types), leading to a postponement of plastic instability. When different deformation modes are sequentially activated, the strain-hardening ability may be regenerated at all such times, leading

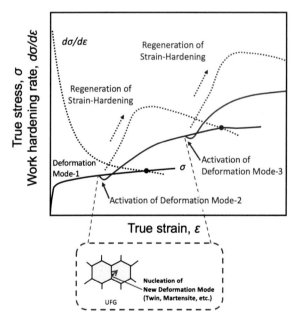

Fig. 1.20 Schematic illustration showing a new strategy, inspired from the results in the bulk nanostructured metals, for managing both high strength and large ductility in metallic materials through sequential nucleation (activation) of different deformation modes and regeneration of strain-hardening rate. The illustration at the bottom illustrates the nucleation of a new deformation mode from a grain boundary in an UFG material, as an example of such a nucleation. Reprinted from Scripta Materialia, vol. 181, Tsuji et al., Strategy for managing both high strength and large ductility in structural materials–sequential nucleation of different deformation modes based on a concept of *plaston*, Copyright (2020), with permission from Elsevier

to high strength and large ductility of the material, as is illustrated in Fig. 1.20.[1] A schematic illustration showing nucleation of new deformation mode from a grain boundary in UFG metals observed above is also shown in Fig. 1.20, as an example of such an activation. The details of the reason why the strain hardening was regenerated by the operation of different deformation modes are still unclear. High-Mn steels and metastable austenitic (Ni–C) steels are originally known to show good balances of strength and ductility even in conventionally coarse grain sizes (Demeri 2013). Their good mechanical properties have been considered due to deformation twinning or deformation-induced martensitic transformation, which has been named TWIP (Grässel et al. 2000; De Cooman et al. 2018) or TRIP (Zackay et al. 1967), respectively. However, the mechanisms of enhanced strain hardening have not yet been exactly clarified even for the conventional TWIP and TRIP phenomena (Luo and Huang 2018).

Another critical point unknown and to be clarified is the activation (nucleation) mechanism of different deformation modes (Li 2007). For designing metallic materials having appropriate chemical compositions and microstructures for sequentially activating different deformation modes and enhancing strain hardening as expressed in Fig. 1.20, we have to figure out the activation mechanism. In order to understand the activation of deformation modes generally, we should consider the energetics and kinetics of the deformation mode under a mechanical loading (stress), as illustrated in Fig. 1.21. Figure 1.21a shows changes in the free energy of the system (material) during the activation of a deformation mode, as a function of the reaction coordinate (collective valuable), i.e., the plastic strain in this case. Figure 1.21b schematically expresses changes of local atomistic structures in a crystal, corresponding to different stages in Fig. 1.21a. Figure 1.21b-1, and b-2 exhibit the status between the initial point (A) and the peak-energy point (B) in Fig. 1.21a, while Fig. 1.21b-3 indicates that between the point (B) and the point (C) in Fig. 1.21a. In Fig. 1.21a, two different free energy curves are drawn, corresponding to the case without stress (black) and the case under a stress (red), respectively. In both cases, the material needs to overcome an energy barrier (ΔG) for activation. When no stress is applied, only thermal activation may help to overcome the barrier (ΔG_0). However, the resultant state (C) would have higher free energy than the initial state (A) owing to lattice defects introduced by the plastic deformation (such as dislocations, surface steps, deformation twins, martensite crystals, and so on), so that the reaction from (A) to (C) via (B) cannot spontaneously happen without mechanical loading (stress). When stress is applied to the material, the activation barrier is reduced from ΔG_0 to ΔG_1, by mechanical activation. (Note that the starting point A is fixed in Fig. 1.21a for both unloading and loading cases.) Correspondingly, it should be noted that the final state (C) becomes

---

[1] Because the initiation of new deformation modes releases elastic energy stored in the material, the flow stress might decrease at that moment, which is reflected in Fig. 1.20. This corresponds to the yield-drop phenomenon universally occurring in recrystallized UFG materials, as discussed above. Such a drop of the stress might be masked in bulky materials having heterogeneities in microstructures, but can appear more obviously in nano-scale materials. Whether such a stress-drop appears on the global stress–strain curve would also depend on the nucleation kinetics of the new deformation mode as well as the degree of heterogeneity of its appearance in the material.

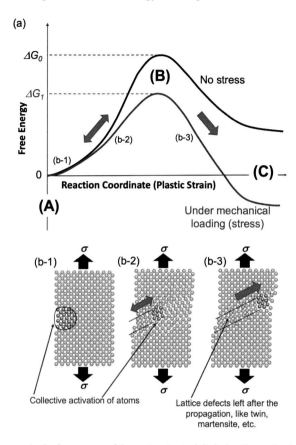

Fig. 1.21 a Changes in the free energy of the system (material) during the nucleation of a deformation mode according to the reaction coordinate (collective valuable), the plastic strain in this case. Two different free energy curves are drawn, corresponding to a case without stress (black) and a case under a stress (red), respectively. **b** Schematic illustrations showing changes of atomistic structures in a crystal, corresponding to different stages of nucleation shown in (**a**). Red atoms correspond to the "*plaston*", i.e., the local defective region collectively activated mechanically and thermally at a singular region in the material. The propagation of the *plaston* brings about plastic deformation and may leave a particular lattice defects, such as stacking faults, deformation twins, martensite, etc. Reprinted from Scripta Materialia, vol. 181, Tsuji et al., Strategy for managing both high strength and large ductility in structural materials–sequential nucleation of different deformation modes based on a concept of *plaston*, Copyright (2020), with permission from Elsevier

lower than the initial state (A) in the case under loading, because of a release of an energy realized by the plastic deformation. This means that a driving force from (A) to (C) arises under a stress.

Let us here consider the change of local atomistic structures during the process in Fig. 1.21b. Here, a perfect crystal without any lattice defects that can be carriers of plastic deformation (like dislocations) is considered. In such a case, we need to nucleate a new deformation mode that leads to a plastic deformation for relaxing the stress. At a singular region with high local stress and/or high energy concentration in the material, such as grain/phase boundary, surface, crack tip, etc., a certain group of atoms would be activated mechanically and thermally, and then form defective zones (drawn by red atoms in Fig. 1.21b). Migration of the local defective zone results in a plastic strain. Between the states (A) and (B) before the energy barrier in Fig. 1.21a, the defective zone may migrate back and forth. After overcoming the barrier (ΔG_1), however, the defective zone migrates in one direction to produce further plastic deformation. The formation and migration of such a local defective zone is the elemental process for the nucleation of a new deformation mode. The propagation of the defective zone may leave a particular defect, such as a dislocation, stacking fault, deformation twin, martensite, rejuvenated glass, and so on, depending on the type of the deformation mode. We would like to call such a localized defective zone of mechanically and thermally excited atoms (expressed by red atoms in Fig. 1.21b) "*plaston*",[2] since it is the essential structure that leads to a plastic strain by its migration. Atomistic structure in dislocation core is one of such localized defective zones. Such structures would be observed at migrating tips of deformation twins, martensite, and shear bands in metallic glasses.

As was mentioned above, enhancing both strength and ductility by deformation twinning and deformation-induced martensitic transformation has been already known as TWIP (Grässel et al. 2000; De Cooman et al. 2018) and TRIP (Zackay et al. 1967), respectively. However, we cannot yet actively control TRIP and TWIP, since we still do not know the critical atomistic process of nucleation for deformation twinning and deformation-induced martensitic transformation. Additionally, we think that it has not yet been clearly proved why global strain hardening of materials is enhanced by deformation twinning (TWIP) and martensitic transformation (TRIP). Dislocation theory is generally powerful to explain plastic deformation and strength of metallic materials, but has a limitation. The dislocation theory is based on elastic fields around dislocations, but the elastic fields of dislocations are obtained assuming Volterra's hollow cylinder (Lee and Nix 2012; Anderson et al. 2017). Therefore, the dislocation theory does not treat discrete atomistic information. As a result, we cannot discuss the nucleation of dislocations. Nucleation and growth of deformation twins

---

[2] Here, the authors need to note that the term "*plaston*" was firstly used by Korbel et al. (1986), Korbel and Martin (1986), Pawełek and Korbel (1990), to the authors' best knowledge. Korbel et al. (1986), Korbel and Martin (1986), Pawełek and Korbel (1990) studied the formation of shear bands in polycrystalline metals. They considered that avalanche-like movement of dislocations in shear banding would promote propagation of local stresses in the way similar to "*soliton*", so that they used the term "*plaston*" for describing the movement of a dislocation group to form a shear band. It should be emphasized, therefore, that the definition of "*plaston*" in the present manuscript and our papers (Tsuji et al. 2020a; Tsuji et al. 2020b) is totally different from that by Korbel et al. (1986), Korbel and Martin (1986), Pawełek and Korbel (1990).

(especially in FCC crystals) and martensite have been often described by the movement of partial or interfacial dislocations and reactions of dislocations, which are, however, just based on geometry. Since conventional metallic materials generally involve large numbers of pre-existing dislocations and easy dislocation sources, as was argued in the Sect. 1.4, it is rather unnecessary to consider nucleation of dislocations in most cases. However, as has been shown above, we need to consider the nucleation of dislocations in nanostructured metals having recrystallized UFG structures. It is also well known that twins and martensite preferentially nucleate from grain boundaries. In the present manuscript, indeed, we have shown that deformation twins, martensite, and $c + a$ dislocations in HCP Mg alloys nucleated from grain boundaries in ultrafine-grained metals (Figs. 1.12, 1.16, 1.17, and 1.18). Those nucleation processes cannot be described by the conventional dislocation theory. We need to understand elementary processes for those phenomena in atomistic scales with their thermodynamics and kinetics, in order to control the sequential nucleation illustrated in Fig. 1.20.

The *plaston* expressed in Fig. 1.21 does describe such atomistic processes. The concept of *plaston* expressed in Fig. 1.21 is useful for various kinds of plastic phenomena, i.e., the nucleation and migration of dislocations (Shinzato et al. 2019), deformation twins (Ogata et al. 2005; Ishii et al. 2016), martensite, disclination, dislocation loops (Zhu et al. 2008, 2007; Du et al. 2016; Li et al. 2018; Sato et al. 2019), disconnections/ledges/steps on grain boundary/interface (Combe et al. 2016; MacKain et al. 2017), vacancy/interstitial clusters, shear transformation in glass (Shimizu et al. 2006, 2007; Zhao et al. 2013; Boioli et al. 2017), and other unknown things in atomistic scales. Recently, we have succeeded in explaining mechanical activation of different deformation modes in recrystallized pure Mg with various average grain sizes based on the concept of *plaston*, according to a simple diagram showing critical stresses for activating different deformation modes from grain boundaries (Zheng et al. 2020b). On the other hand, Mao (2021) has experimentally quantified the activation energy for deformation-induced martensitic transformation in Fe–23Ni–3.55Mn alloy. Such accumulation of experimental analyses and computer simulations in atomistic scale considering *plaston* would deepen the understanding of activation and migration properties of the localized defective zone (*plaston*) that produces plasticity. Different kinds of *plaston* would have different activation energies and different dependencies on stress and temperature. Once we figure out those properties, we would be able to design optimized material with appropriate chemical compositions and microstructures and to realize optimized processes at appropriate temperature and strain rate for controlling the activation of *plaston*. The concept of *plaston* would be also useful for considering fatigue and fracture behavior of materials, since regions near crack-tips are typical singular points of stress. Then it would become possible not only to overcome the strength-ductility trade-off, but also to make less deformable materials plastic.

1.5 Conclusions

Nowadays, extremely high strength is often required for structural materials, but managing both high strength and large ductility has been a challenge in any kind of material including metals and alloys. Bulk nanostructured metals composed of ultrafine grains with average grain sizes smaller than 1 μm exhibit very high strength compared to their coarse-grained counterparts, but bulk nanostructured metals generally show limited uniform tensile elongation of a few % due to the early plastic instability. On the other hand, the present authors have found several exceptions of bulk nanostructured alloys with recrystallized structures that can overcome the strength-ductility trade-off. In such bulk nanostructured alloys, unexpected deformation modes were activated because of the scarcity of dislocations and dislocation sources in each recrystallized ultrafine grain. Concrete examples of unexpected deformation modes activated were $<c + a>$ dislocations in a Mg alloy, nano deformation twins in high-Mn austenitic steels, and deformation-induced martensite in metastable austenitic steels. All those newly activated deformation modes seemed to nucleate from grain boundaries in the nanostructured metals, and they enhanced strain hardening of the materials (probably due to interactions between different deformation modes including normal dislocations), leading to the postponement of plastic instability and excellent mechanical properties managing high strength and large tensile ductility. Based on such experimental results in bulk nanostructured metals, the authors have proposed a new strategy of sequential nucleation of different deformation modes for realizing advanced structural materials showing excellent mechanical properties and a new concept of *plaston* for understanding the nucleation process of deformation modes under mechanical loading. Understanding of the *plaston* concept would make it possible to design advanced structural materials that control the activation of various deformation modes in appropriate timing in deformation and would give a fundamental guiding principle for managing both high strength and high ductility/toughness.

Acknowledgements This work was supported by the Elements Strategy Initiative for Structural Materials (ESISM) of MEXT (Grant number JPMXP0112101000), in part by JSPS KAKENHI (Grant numbers JP15H05767, JP20H00306), and by JST CREST (Grant number JPMJCR1994).

References

Altan BS (2006) Severe plastic deformation toward bulk production of nanostructured materials. NOVA Science Publishers, NY

Anderson PM, Hirth JP, Lothe J (2017) Theory of dislocations, 3rd edn. Cambridge University Press, Cambridge

Azushima A, Kopp R, Korhonen A, Yang DY, Micari F, Lahoti GD, Groche P, Yanagimoto J, Tsuji N, Rosochowski A, Yanagida A (2008) SIRP Ann 57:716–735

Bai Y, Kitamura H, Gao S, Tian Y, Park N, Park M, Adachi H, Shibata A, Sato M, Murayama M, Tsuji N (2021) Unique transition of yielding mechanism and unexpected activation of deformation twinning in ultrafine grained Fe–31Mn–3Al–3Si alloy, Sci Rep 11:15870

Bai Y, Momotani Y, Chen MC, Shibata A, Tsuji N (2016) Mater Sci Eng A 651:935–944

Bai Y, Tian Y, Gao S, Shibata A, Tsuji N (2018) J Mater Res 32:4592–4604

Bei H, Shim S, Pharr GM, George EP (2008) Acta Mater 56:4762–4770

Boioli F, Albaret T, Rodney D (2017) Phys Rev E 95:033005

Chen S (2015) PhD thesis. Kyoto University

Chen S, Shibata A, Gao S, Tsuji N (2014) Mater Trans 55:223–226

Chen ZMT, Okamoto NL, Demura M, Inui H (2016) Scripta Mater 121:28–31

Chong Y, Deng G, Gao S, Yi J, Shibata A, Tsuji N (2019) Yielding nature and Hall-Petch relationship in Ti–6Al–4V alloy with fully equiaxed and bimodal microstructures. Scripta Mater 172:77–82

Combe N, Mompiou F, Legros M (2016) Phys Rev B 024109

Cohen J, Weertman J (1963) A dislocation model for twinning in fcc metals. Acta Metall 11:996–998

De Cooman BC, Estrin Y, Kim SK (2018) Twinning-induced plasticity (TWIP) steels. Acta Mater 142:283–362

Copper N, Fang TH, Li WL, Tao NR, Lu K (2011) Science 331:1587–1590

Demeri MY (2013) Advanced high-strength steels: science, technology, and applications. ASM International, Materials Park, Ohio

Du J-P, Wang Y-J, Lo Y-C, Wan L, Ogata S (2016) Phys Rev B 94:104110

Fujimura K, Kishida K, Tanaka K, Inui H, Symp MRS (2011) Proc 1295:201–206

Gao S, Bai Y, Zheng R, Tian Y, Mao W, Shibata A, Tsuji N (2019) Mechanism of huge Luders deformation in ultrafine grained austenitic stainless steel. Scripta Mater 159:28–32

Gao S, Chen MC, Chen S, Kamikawa N, Shibata A, Tsuji N (2014a) Yielding behavior and its effect on uniform elongation of fine grained IF steel. Mater Trans 55:73–77

Gao S, Chen M, Chen S, Kamikawa N, Shibata A, Tsuji N (2014b) Mater Trans 55:73–77

Grässel O, Kruger L, Frommeyer G, Meyer L (2000) Int J Plast 16:1391–1409

Greer JR, Oliver WC, Nix WD (2005) Acta Mater 53:1821–1830

Hall EO (1951) Proc Phys Soc London B 64:742–747

Hall EO (1970) Yield point phenomena in metals & alloys. Plenum Press, New York

He BB, Hu B, Yen HW, Cheng GJ, Wang ZK, Luo HW, Huang MX (2017) Science 357:1029–1032

Higashi M, Momono S, Kishida K, Okamoto NL, Inui H (2018) Acta Mater 161:161–170

Huang X, Hansen N, Tsuji N (2006) Hardening by annealing and softening by deformation in nanostructured metals. Science 312:249–251

Hung C-Y, Bai Y, Tsuji N, Murayama M (2021a) Grain size altering yielding mechanisms in ultrafine grained highustenitic steel: Advanced TEM investigations. J Mater Sci Tech 86:192–203

Hung C-Y, Bai Y, Shimokawa T, Tsuji N, Murayama M (2021b) A correlation between grain boundary character and deformation twin nucleation mechanism in coarse-grained high-Mn austenitic steel. Sci Rep 11:8468

Hung C-Y, Shimokawa T, Bai Y, Tsuji N, Murayama M (2021c) Investigating the dislocation reactions on $\Sigma 3\{111\}$ twin boundary during deformation twin nucleation process in an ultrafine-grained high-manganese austenitic steel. Sci Rep 11:19298

Hwang S, Park M, Bai Y, Shibata A, Mao W, Adachi H, Sato M, Tsuji N (2021) Mesoscopic nature of serration behavior in high-Mn austenitic steel. Acta Mater 205:116543

Inoue A, Kishida K, Inui H, Hagihara K, Symp MRS (2013) Proc 1516:151–156

Ishii A, Li J, Ogata S (2016) Int J Plast 82:32–43

Jennings AT, Weinberger CR, Lee SW, Aitken ZH, Meza L, Greer JR (2013) Acta Mater 61:2244–2259

Kamikawa N, Huang X, Tsuji N, Hansen N (2009) Strengthening mechanisms in nanostructured high purity aluminum deformed to high strains and annealed. Acta Mater 157:4198–4208

Kim S-H, Kim H, Kim NJ (2015) Nature 518:77–79

Kishida K, Kim JG, Nagae T, Inui H (2020) Experimental evaluation of critical resolved shear stress for first-prder pyramidal c + a slip in commercially pure Ti by micropillar compression method. Acta Mater 196:168–174

Kishida K, Maruyama T, Matsunoshita H, Fukuyama T, Inui H (2018) Acta Mater 159:416–428

Kitamura H (2017) Master thesis. Kyoto University

Kraft O, Gruber PA, Mönig R, Weygand D (2010) Annu Rev Mater Res 40:293–317

Korbel A (1986) A real nature of shear bands-plastons. Archiwum Hutnictwa 31:33–41

Korbel A, Martin P (1986) Microscopic versus macroscopic aspect of shear bands deformation. Acta Metall 34:1905–1909

Lee SW, Nix WD (2012) Philos Mag 92:1238–1260

Lei Z, Liu X, Wu Y, Wang H, Jiang S, Wang S, Hui X, Wu Y, Gault B, Kontis P, Raabe D, Gu L, Zhang Q, Chen H, Wang H, Liu J, An K, Zeng Q, Nieh TG, Lu Z (2018) Nature 563:546–550

Li J (2007) MRS Bull 32:151–159

Li H, Gao S, Tomota Y, Ii S, Tsuji N, Ohmura T (2021) Mechanical response of dislocation interaction with grain boundary in ultrafine-grained interstitial-free steel. Acta Mater 206:116621

Li Q-J, Xu B, Hara S, Li J, Ma E (2018) Acta Mater 145:19–29

Li Z, Pradeep KG, Deng Y, Raabe D, Tasan CC (2016) Nature 534:227–230

Lu L, Chen X, Huang X, Lu K (2009) Science 323:607–610

Liu L, Ding Q, Zhong Y, Zou J, Wu J, Chiu YL, Li J, Zhang Z, Yu Q, Shen Z (2018) Mater Today 21:354–361

Liu G, Zhang GJ, Jiang F, Ding XD, Sun YJ, Sun J, Ma E (2013) Nat Mater 12:344–350

Luo ZC, Huang MX (2018) Scripta Mater 142:28–31

Ma E, Zhu T (2017) Mater Today 20:323–331

MacKain O, Cottura M, Rodney D, Clouet E (2017) Phys Rev B 134102

Mao W (2021) PhD Thesis. Kyoto University

Mahajan S, Chin G (1975) Comments on deformation twinning is silver-and copper-alloy crystals. Scr Metall 9:815–817

Miura S, Takamura J, Narita N (1968) Trans JIM 9:555–561

Morris JW Jr (2008) ISIJ Int 48:1063–1070

Nakatsuka S, Kishida K, Inui H (2015) MRS Symp Proc 1760, mrsf14-1760-yy05-09

Ogata S, Li J, Hirosaki N, Shibutani Y (2004) Phys Rev B 70:104104

Ogata S, Li J, Yip S (2002) Science 298:807–811

Ogata S, Li J, Yip S (2005) Phys Rev B 71:224102

Okamoto NL, Fujimoto S, Kambara Y, Kawamura M, Chen ZMT, Matsunoshita H, Tanaka K, Inui H, George EP (2016) Sci Rep 6:35863

Okamoto NL, Kashioka D, Inomoto M, Inui H, Takebayashi H, Yamaguchi S (2013) Scripta Mater 69:307–310

Okamoto NL, Inomoto M, Adachi H, Takebayashi H, Inui H (2014) Acta Mater 65:229–239

Parthasarathy TA, Rao SI, Dimiduk DM, Uchic MD, Trinkle DR (2007) Scripta Mater 56:313–316

Pawełek A, Korbel A (1990) Soliton-like behavior of a moving dislocation group. Phil Mag B 61:829–842

Petch NJ (1953) J Iron Steel Inst 174:25–28

Saha R, Ueji R, Tsuji N (2013) Fully recrystallized nanostructure fabricated without severe plastic deformation in high-Mn austenitic steel. Scr Mater 68:813–816

Saito Y, Tsuji N, Utsunomiya H, Sakai T, Hong RG (1998) Scripta Mater 39:1221–1227

Saito Y, Utsunomiya H, Tsuji N, Sakai T (1999) Acta Mater 47:579–583

Sato Y, Shinzato S, Ohmura T, Ogata S (2019) Int J Plast 121:280–292

Shimizu F, Ogata S, Li J (2006) Acta Mater 54:4293–4298

Shimizu F, Ogata S, Li J (2007) Mater Trans 48:2923–2927

Shinzato S, Wakeda M, Ogata S (2019) Int J Plast 122:319–337

Sudharshan Phani P, Johanns KE, George EP, Pharr GM (2013) Acta Mater 61:2489–2499

Sun W, Zhu Y, Marceau R, Wang L, Zhang Q, Gao X, Hutchinson C (2019) Science 363:972–975

Surya EE-D, Kalidindi R, Doherty RD (1999) Metall Mater Trans A 30:1223–1233

Takechi H (1994) ISIJ Int 34:1–8

Terada D, Inoue M, Kitahara H, Tsuji N (2008) Change in mechanical properties and microstructures of ARB processed Ti during annealing. Mater Trans 49:41–46

Tian YZ, Bai Y, Chen MC, Shibata A, Terada D, Tsuji N (2014) Metall Mater Trans A 45:5300–5301

Tian YZ, Gao S, Zhao LJ, Lu S, Pippan R, Zhang ZF, Tsuji N (2018a) Remarkable transitions of yield behavior and Lüders deformation in pure Cu by changing grain sizes. Scripta Mater 142:88–91

Tian YZ, Xiong T, Zheng SJ, Bai Y, Freudenberger J, Pippan R, Zhang ZF, Tsuji N (2018b) Materialia 3:162–168

Tian YZ, Gao S, Zheng RX, Wang JH, Ren YP, Pan HC, Qin GW, Zhang ZF, Tsuji N (2020a) Two-stage Hall-Petch relationship in Cu with recrystallized structure. J Mater Sci Tech 48:31–35

Tian YZ, Ren YP, Gao S, Zheng RX, Wang JH, Pan HC, Zhang ZF, Tsuji N, Qin GW (2020b) Two-stage Hall-Petch relationship in Cu with recrystallized structure. J Mater Sci Tech 31–35

Tian YZ, Zhao LJ, Chen S, Shibata A, Zhang ZF, Tsuji N (2015) Sci Rep 5:16707

Tian YZ, Zhao LJ, Park N, Liu R, Zhang P, Zhang ZJ, Shibata A, Zhang ZF, Tsuji N (2016) Acta Mater 110:61–72

Tong Y, Zhao YL, Hu A, Lu K, Cai JX, Kai JJ, Liu Y, Yang T, Wei J, Liu CT, Han XD, Jiao ZB, Chen D (2018) Science 362:933–937

Tsuji N (2002) Ultrafine grained steels. Tetsu-to-Haganè 88:359–369

Tsuji N (2007) Unique mechanical properties of nano-structured metals. J Nanosci Nanotechnol 7:3765–3770

Tsuji N, Gholizadeh R, Ueji R, Kamikawa N, Zhao L, Tian Y, Bai Y, Shibata A (2019) Formation mechanism of ultrafine grained microstructures: various possibilities for fabricating bulk nanostructured metals and alloys. Mater Trans 60:1518–1532

Tsuji N, Hansen N, Huang X, Godfrey A (2017) Personal communications

Tsuji N, Kamikawa N, Ueji R, Takata N, Koyama H, Terada D (2008) Managing both strength and ductility in ultrafine grained steels. ISIJ Int 48:1114–1121

Tsuji N, Ito Y, Saito Y, Minamino Y (2002) Strength and ductility of ultrafine grained aluminum and iron produced by ARB and annealing. Scripta Mater 47:893–899

Tsuji N, Ogata S, Inui H, Tanaka I, Kishida K, Gao S, Mao W, Bai Y, Zheng R, Du JP (2020) Strategy for managing both high strength and large ductility in structural materials—sequential nucleation of different deformation modes based in a concept of plaston. Scripta Mater 181:35–42

Tsuji N, Ogata S, Inui H, Tanaka I, Kishida K, Gao S, Mao W, Bai Y, Zheng R, Du JP (2021) Corrigendum to 'Strategy for managing both high strength and large ductility in structural materials—sequential nucleation of different deformation modes based on a concept of plaston. Scripta Mater 181:35–42; SMM 13102, Scripta Mater 196:113755

Tsuji N, Saito Y, Lee SH, Minamino Y (2003) Adv Eng Mater 5:338–344

Uchic MD, Dimiduk DM, Florando JN, Nix WD (2004) Science 305:986–989

Uchic MD, Shade PA, Dimiduk DM (2009) Annu Rev Mater Res 39:361–386

Umemoto M, Owen WS (1974) Metall Trans 5:2041–2046

Venables JA (1961) Deformation twinning in face-centered-cubic metals. Philos Mag 6:396–397

Wagoner RH, Chenot JL (1997) Fundamentals of metal formong. Wiley, New York

Wang Y, Chen M, Zhou F, Ma E (2002) Nature 419:912

Wei Y, Li Y, Zhu L, Liu Y, Lei X, Wang G, Wu Y, Mi Z, Liu J, Wang H, Gao H (2014) Nat Comm 5:3580

Wu X, Jiang P, Chen L, Yuan F, Zhu YT (2014) Proc Natl Acad Sci 111:7197–7201

Wu X, Yang M, Yuan F, Wu G, Wei Y, Huang X, Zhu Y (2015) Proc Natl Acad Sci 112:14501–14505

Yang M, Yan D, Yuan F, Jiang P, Ma E, Wu X (2018) Proc Natl Acad Sci 115:7224–7229

Yoshinaga H, Horiuchi R (1964) Trans JIM 5:14–21

Yoshida S, Bhattacharjee T, Bai Y, Tsuji N (2017) Friction stress and Hall-Petch relationship in CoCrNi equi-atomic medium entropy alloy processed by severe plastic deformation and subsequent annealing. Scripta Mater 134:33–36

Yoshida S, Ikeuchi T, Bhattacharjee T, Bai Y, Shibata A, Tsuji N (2019) Effect of elemental combination on friction stress and Hall-Petch relationship in face-centered cubic high/medium entropy alloys. Acta Mater 171:201–215

Zackay VF, Parker ER, Fahr D, Bush R (1967) Trans ASM 60:252–259

Zhang B, Chong Y, Zheng R, Bai Y, Gholizadeh R, Huang M, Wang D, Sun Q, Wang Y, Tsuji N (2020) Enhanced mechanical properties in beta-Ti alloy aged from recrystallized ultrafine beta grains. Mater Des 109017

Zhang J, Beyerlein IJ, Han W (2019) Phys Rev Lett 122:255501

Zhang J, Kishida K, Inui H (2017) Int J Plast 92:45–56

Zhao P, Li J, Wang Y (2013) Int J Plast 40:1–22

Zhao YH, Liao XZ, Cheng S, Ma E, Zhu YT (2006) Adv Mater 18:2280–2283

Zheng RX, Bhattacharjee T, Shibata A, Sasaki T, Hono K, Joshi M, Tsuji N (2017) Simultaneously enhanced strength and ductility of Mg–Zn–Zr–Ca alloy with fully recrystallized ultrafine grained structures. Scripta Mater 131:1–5

Zheng RX, Bhattacharjee T, Gao S, Gong W, Shibata A, Sasaki T, Hono K, Tsuji N (2019) Change of deformation mechanisms leading to high strength and large ductility in Mg–Zn–Zr–Ca alloy with fully recrystallized ultrafine grained microstructures. Sci Rep 9:11702

Zheng R, Du JP, Gao S, Somekawa H, Ogata S, Tsuji N (2020a) Transition of dominant deformation mode in bulk polycrystalline pure Mg by ultra-grain refinement down to sub-micrometer. Acta Mater 198:35–46

Zheng R, Du JP, Gao S, Somekawa H, Ogata S, Tsuji N (2020b) Transition of dominant deformation mode in bulk polycrystalline pure Mg by ultyra-grain refinement down to sub-micrometer. Acta Mater 198:35–46

Zhu T, Li J (2010) Prog Mater Sci 55:710–757

Zhu T, Li J, Ogata S, Yip S (2009) MRS Bull 34:167–172

Zhu T, Li J, Samanta A, Kim HG, Suresh S (2007) Proc Natl Acad Sci 104:3031–3036

Zhu T, Li J, Samanta A, Leach A, Gall K (2008) Phys Rev Lett 100:025502-1-25504

Part II
Simulation of *Plaston* and *Plaston* Induced Phenomena

Chapter 2
Free-energy-based Atomistic Study of Nucleation Kinetics and Thermodynamics of Defects in Metals; Plastic Strain Carrier "Plaston"

Shigenobu Ogata

2.1 Introduction

Unlike elastic deformation, which tends to occur with a broader and more uniform strain distribution, plastic deformation proceeds in a more localized manner. It is realized through the nucleation and migration of local atomistic defects, including dislocations, disconnections, disclinations, vacancies/impurities/interstitials in crystals, and shear transformation zones in glasses. These defects are collectively called "plastons" in this book, as described in the previous section. The defects function as "carriers" of plastic strain, releasing the elastic tension/compression (reducing internal elastic strain energy) by their motion in plastically deformable materials, such as metals. These defect activities eventually induce changes in the material structure and texture (e.g., phase transformation, twinning, stacking fault formation, crack propagation and blunting, surface morphology change, grain growth and rotation, and glass relaxation). In plasticity and its dynamics, temperature- and stress-dependent deformation kinetics (e.g., the strain rate and its temperature and stress dependencies) are crucial factors. These factors often shift the dominant deformation process in a material by changing the kinetics of the available defect activities. For example, the dominant deformation process in creep transits from diffusive (atomic diffusion and grain boundary (GB) migration and sliding) to more displacive (dislocation glide) with increasing stress and/or decreasing temperature (Wang et al. 2011). Therefore, studying all the available defect activities and the corresponding kinetics under different temperatures and stress conditions is vital for understanding, and thus controlling, plasticity dynamics.

S. Ogata (✉)
Department of Mechanical Science and Bioengineering,
Graduate School of Engineering Science, Osaka University, Osaka 560-8531, Japan
e-mail: ogata@me.es.osaka-u.ac.jp

Center for Elements Strategy Initiative for Structural Materials (ESISM), Kyoto University,
Sakyo-ku, Kyoto 606-8501, Japan

© The Author(s) 2022   37
I. Tanaka et al. (eds.), *The Plaston Concept*,
https://doi.org/10.1007/978-981-16-7715-1_2

The question arises as to what can be done for understanding and controlling plasticity dynamics. Uncovering the free-energy landscape with an appropriate corrective variable (CV) for the available defect activities is a potential answer. The free-energy landscape directly provides kinetic-related information about defects via the activation energy, which is characterized as the difference in free energy between the local equilibrium and saddle states of the considered defect activity. Additionally, the temperature and stress dependencies of kinetics are naturally described from those of free energy via parameters such as the activation parameters of activation volume and activation entropy. Furthermore, thermodynamics are fully described according to the free-energy difference between local equilibrium states. Hence, once the free-energy landscape covering the possible (accessible) defect activities is understood, the plasticity dynamics of the materials can be completely defined and predicted for any temperature and stress conditions. Atomistic modeling methods based on a reliable energy description, such as density functional theory and sophisticated interatomic interaction, are promising tools for elucidating the free-energy landscape of the defect activities, because these activities are usually in atomic scale. Fortunately, substantial advances in atomistic modeling methods have recently reported to elucidate the free-energy landscape.

Notably, the free-energy landscape exhibits a multiscale nature in many cases, as mentioned in classical nucleation theory. For example, studying a dislocation nucleation process from a nucleation site in a material according to the free-energy concept can reveal a saddle point configuration at a loop size during the dislocation loop expansion process. However, a careful examination of a segment of the dislocation line and its motion at the atomic scale can reveal "local" saddle points attributable to the individual segment motion successively overcoming the Peierls potential barrier. All these saddle points contribute to the kinetics of the dislocation loop nucleation; however, the former saddle point defines the activation free-energy barrier of the entire loop nucleation process.

In this section, atomistic modeling studies on plaston kinetics and thermodynamics, such as the nucleation of deformation twins and the heterogeneous and homogeneous nucleation of dislocations, are introduced from the free-energy standpoint.

2.2 Shuffling Dominant $\{10\bar{1}2\}\langle10\bar{1}\bar{1}\rangle$ Deformation Twinning in Hexagonal Close-Packed Magnesium (Ishii et al. 2016)

In deformation twinning (DT) (Christian Mahajan 1995), a crystal is transformed into a mirrored configuration with transformation strain ε_{final}, which is as important deformation mode as dislocation. Although many atomistic DT simulation and experiments have been performed, the DT nucleation pathway and kinetics are still unclear. These issues remain controversial issues in the study of plasticity. The nucleation pathway and kinetics must be dominated by twin boundary nucleation and migra-

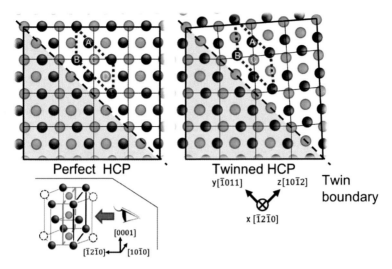

Fig. 2.1 Atomic arrangements of perfect and twinned hexagonal close-packed (HCP) configurations viewed from $[1\bar{2}10]$ (Ishii et al. 2016). The four-atom supercell shape is depicted by the red dotted line

tion, which can be driven by local atomic-scale shear deformation. In many cases, interatomic-layer sliding can realize atomic-scale local shear deformation. However, in some cases, for instance, the following hexagonal close-packed (HCP) case, more complicated atomic motion, such as nonaffine atomic "shuffling", is occurs because of its lower activation Gibbs free energy.

Figure 2.1 presents perfect ($\lambda = 0$) and twinned ($\lambda = 1$) HCP atomic structures corresponding to $\langle 1\bar{2}10 \rangle$, at $\sigma = 0$. The four-atom supercell ($M = 4$) includes atoms A, B, C, and D is the minimum lattice correspondence pattern unit required to render the atomic arrangements during $\{10\bar{1}2\}\langle 10\bar{1}\bar{1}\rangle$ twinning shear deformation (Li and Ma 2009; Wang et al. 2013). The deformation unit has internal degrees of freedom geometrically independent of lattice strain. Importantly, external stress cannot independently control the internal degrees of freedom, referred to as the so called nonaffine atomic "shuffling," from the lattice strain because the internal degrees of freedom are only "slaves" of the affine lattice strain. However, actual deformation transpires at finite temperatures. Thermal fluctuations are induced by thermal energy. Thermal energy can independently disturb the internal degrees of freedom from the lattice strain (i.e., free the degrees of freedom from slavery) and excite them. Therefore, it can enable to attain instability point (saddle point) on the Gibbs free-energy landscape to not be taken only by the application of stress, where the soft mode is mostly along the direction of the nonaffine atomic shuffling. Thus, the method for applying stress at zero temperature may not discover these instability points. A Gibbs free-energy landscape in a space spanned by lattice strain and shuffling degrees of freedom can resolve this issue, thereby allowing the estimation of minimum energy pathway (MEP) and corresponding activation Gibbs free energy considering the nonaffine atomic shuffling in addition to affine lattice strain.

A scalar for representing the shuffling degree of freedom (Ishii et al. 2016) is defined as

$$I \equiv (s - s^{\text{ini}})^T \mathbf{H}_0^T \mathbf{H}_0 (s^{\text{fin}} - s^{\text{ini}})/M \qquad (2.1)$$

where s^{ini} and s^{fin} indicate the internal coordinates of the labeled atoms before and after deformation ($\lambda = 0$ and 1), respectively, for a given σ. Note that, the s differences in (2.1) corresponds to the changes in internal coordinates with no periodic boundary condition (PBC) wraparound. The unit of I is Å^2. When the deformation is complete, I takes the meaning of mean square nonaffine displacements (MSDs). The crystal is transformed from a reference configuration to another configuration with strain $\varepsilon_{\text{final}}$. The supercell that describes the deformation can be taken as an irreducible lattice correspondence pattern, which can be greater than the host lattice primitive cell. The deformation can be atomistically express as

$$\mathbf{x}_m(\lambda) = \mathbf{H}(\lambda)\mathbf{s}_m(\lambda), \quad \mathbf{H}(\lambda) = \mathbf{R}(\lambda)(\mathbf{I} + 2\varepsilon(\lambda))^{1/2} \mathbf{H}_0, \qquad (2.2)$$

where $\mathbf{x}_m$ denotes the Cartesian position, λ denotes the reaction progress variable (scalar), and $\mathbf{s}_m = [s_{m1}; s_{m2}; s_{m3}] \in [0, 1)$ refers to the reduced coordinate vector of atom m under PBC. Further, $\mathbf{H} = [\mathbf{h}_1\mathbf{h}_2\mathbf{h}_3]$ is a 3×3 matrix, where $\mathbf{h}_1$, $\mathbf{h}_2$, and $\mathbf{h}_3$ corresponding to the three edge vectors of the supercell and $m = 1, \ldots, M$ is the atom index in the supercell. $(\mathbf{I} + 2\varepsilon(\lambda))^{1/2}$ and $\mathbf{R}(\lambda)$ are the equation components corresponding to the irrotational and rotational parts of the deformation gradient, respectively, where $\varepsilon(\lambda)$ denotes the Lagrangian strain with respect to the initial configuration.

The MEP with the least M can then be computed based on ab initio first-principles computation at constant external stress σ, yielding the activation Gibbs energy $G(\lambda, \sigma)$ versus reaction coordinate λ, that is parametrized by σ. An algorithm, such as the nudged elastic band (NEB) method (Jonsson et al. 1998), can be employed to obtain the MEP and fix the saddle point on the MEP:

$$Q(\sigma) \equiv G(\lambda^*, \sigma) - G(0, \sigma) \qquad (2.3)$$

on the joint $\varepsilon \otimes \mathbf{s}$ space (Sheppard et al. 2012), where $\lambda = 0$ and 1 denote the state before and after deformation, respectively, and λ^* denotes a saddle point, at constant external stress σ. The Gibbs free-energy landscape can be numerically estimated using ab initio first-principles computation by changing ε and $\mathbf{s}$.

$$G(\varepsilon, s, \sigma) \equiv U(\varepsilon, s) - W(\varepsilon, \sigma), \qquad (2.4)$$

where $W(\varepsilon, \sigma)$ is the work performed by constant external Cauchy stress σ (Wang et al. 1995):

$$W(\varepsilon, \boldsymbol{\sigma}) \equiv \int_0^1 dl \det |\mathbf{J}(\eta = l\varepsilon)\mathbf{H}_0| \times \mathrm{Tr}\left[\mathbf{J}^{-1}(\eta = l\varepsilon)\boldsymbol{\sigma}\mathbf{J}^{-T}(\eta = l\varepsilon)\varepsilon\right], \quad (2.5)$$

where $\eta = l\varepsilon = 1/2(\mathbf{J}^{\mathsf{T}}\mathbf{J} - \mathbf{I})$ denotes the Lagrangian strain tensor, and $\mathbf{J}$ denotes the corresponding deformation gradient tensor,

$$\mathbf{J} = \mathbf{R}(\mathbf{I} + 2\eta)^{1/2} \qquad (2.6)$$

where $\mathbf{R}$ is an additional rotation matrix $\mathbf{R}^{\mathsf{T}}\mathbf{R} = \mathbf{I}$ that is completely defined when the transformation coordinate frame convention is selected. Although $G(\varepsilon, s, \boldsymbol{\sigma})$ is now defined using (2.4), its direct visualization is difficult because the $\varepsilon \otimes s$ space is $3M + 6$-dimensional. Therefore, (2.1) can be used to aid visualization. We can uniquely compute $G(\gamma, I, \boldsymbol{\sigma})$ by implementing energy minimization to all degrees of freedom of the supercell system other than γ and I:

$$G(\gamma, I, \boldsymbol{\sigma}) \equiv \min_{\varepsilon \in \gamma, s \in I} G(\varepsilon, s, \boldsymbol{\sigma}) \qquad (2.7)$$

Figure 2.2 presents the Gibbs free-energy landscapes $\Delta G(\gamma, I)$ obtained at different external Cauchy shear stresses ($\sigma_{yz} = \sigma_{zy} = 0.0$, 1.0, and 2.0 GPa in the twinning direction). The red curves on the Gibbs free-energy landscapes indicate the MEPs from the original to the twinned configuration under these external shear stresses, which were determined using the NEB method. The change in the external shear stress shifts the equilibrium state before and after the twinning as well as the saddle point. Notably, the saddle point is located at a point of finite I and the MEP parallels the I-axis more closely than it parallels the γ axis, suggesting that I dominates the DT process. In this case, merely achieving the shear strain γ is insufficient to overcome the activation Gibbs free-energy barrier. Figure 2.4a shows the Gibbs free-energy profile along the MEP at different external shear stresses. The change in the Gibbs free-energy barrier with respect to the external shear stress are presented in Fig. 2.4b. To confirm the above discussion, two NEB calculations were independently performed: (1) with respect to internal atomic configuration I, where the supercell shape is relaxed under the predefined external stress (I-control NEB) and (2) with respect to γ, where the internal atomic configuration s is relaxed for each supercell frame shape (γ-control NEB). In Fig. 2.4b, the Gibbs free-energy barriers of the two calculations are compared; results indicates that the Gibbs free-energy barrier obtained using I-control NEB matches that obtained using the two-dimensional Gibbs free-energy landscape. Alternatively, the Gibbs free-energy barrier using γ-control NEB is substantially greater that obtained using I-control NEB. These results clearly indicate that DT corresponding to an I-dominant (nonaffine-displacement dominant) deformation and not γ-dominant. Hence, the twinned structure can be generated first without producing local shear strain because the phonons can toggle the "internal cog" at finite temperatures. After flipping the internal cog, the local γ can later spontaneously relax along the twinning configuration. Because these processes actually occur almost simultaneously in the DT process, it is impossible to use DT atomic motion observations to determine whether I- or γ-dominant. The

Fig. 2.2 Gibbs free-energy landscapes $\Delta G(\gamma, I)$ at $\sigma_{yz} = \sigma_{zy}$ = values of **a** 0.0, **b** 1.0, and **c** 2.0 GPa. The red-dashed curves denote the minimum energy paths (MEPs) of the deformation twinning processes (Ishii et al. 2016)

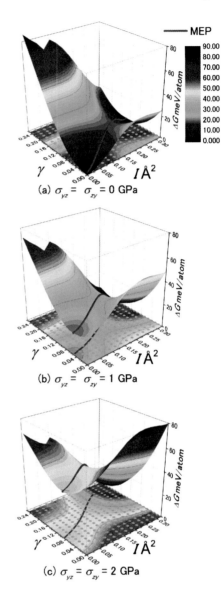

(a) $\sigma_{yz} = \sigma_{zy} = 0$ GPa

(b) $\sigma_{yz} = \sigma_{zy} = 1$ GPa

(c) $\sigma_{yz} = \sigma_{zy} = 2$ GPa

Gibbs free-energy landscape analysis is required to obtain insights into the fundamental mechanism. Figure 2.3 shows the atomic position and supercell shape change along the MEP under the stress-free conditions. A uniform supercell shape change (shear strain) with a staggered rotation of A–B and C–D bonds is clearly observed. Intuitively, the latter bond rotation behavior is hard to archive only using an external shear stress along the DT direction. Figure 2.4b demonstrates that a very high critical external shear stress of ∼3.0 GPa is necessary to realize the DT by γ-control at

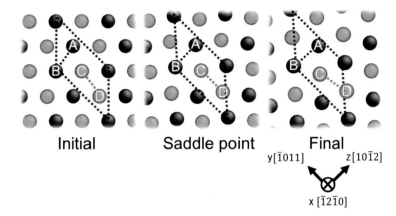

Fig. 2.3 Atomic position and supercell shape change along the MEP (Ishii et al. 2016). The green and black atoms are located in different atomic layers

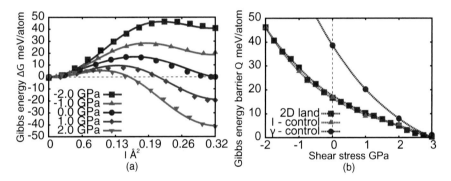

Fig. 2.4 a Gibbs free-energy profile along the MEP and **b** change in the energy barrier along the MEP with respect to change in applied external shear stress. Figure (b) indicates the DT process is I-dominant, because the Gibbs energy barrier of the I-controlled NEB reproduces the Gibbs free-energy barriers obtained using the two-dimensional Gibbs free-energy landscape (2D-land) (Ishii et al. 2016)

athermal condition, while considerably less lower external shear stress is required at finite temperatures because the thermal energy activates the system toward the I direction.

Hence, the free-energy landscape with appropriate CVs, such as I and γ, successfully describes the hidden saddle point of the shuffling-dominant deformation twinning of HCP. Notably, deformation using shuffling should not be specific to HCP DT, it should be omnipresent and thus should be found in shear deformations in FCC and BCC metals, glasses, and ceramics with complicated crystal structures.

2.3   Dislocation Nucleation from GBs (Du et al. 2016)

The dislocation nucleation from interfacial defects dominates the plastic deformation of materials with limited small volumes, which may have a limited number and/or activities of plastic deformation carriers. For instance, the plastic deformation of nanocrystalline metals exhibiting high strength, is led by dislocation nucleation from GBs (Wang et al. 2011) at low temperatures and high strain rates; this dislocation nucleation is activated at a stress higher than that necessary for the usual dislocation motion. Molecular dynamics (MD) is among the best tools for examining the dislocation nucleation from GBs because the nucleation event is atomistic, thus enabling a close examination of the details. However, in MD simulations, the typical strain rate $\sim 10^6 \ s^{-1}$ substantially differs from that in the experiments ($\sim 10^{-3} \ s^{-1}$) because of the MD simulations' limited timescale. Therefore, to study the temperature and strain rate sensitivities, accelerated MD methods, such as adaptive-boost MD (ABMD) (Ishii et al. 2012, 2013), can be used, which is also free-energy-based atomistic modeling. The benefit of using ABMD is that it enables the direct computation of not only stress-dependent but also temperature-dependent activation free energy.

The ABMD method was employed to study the dislocation nucleation event from a GB, $\Sigma = 9\langle 110 \rangle \{221\}$ symmetric tilt grain boundary in FCC Cu, under conditions of lower external stress and temperatures, where regular MD is not applicable because of the longer incubation time for dislocation nucleation. The ABMD directly estimate the nucleation frequency (incubation time). Based on the nucleation frequency, the free-energy barrier and activation enthalpy and entropy, can be computed using the transition-state theory. A bias potential (boost potential) is added by the ABMD method to the original potential, and the bias potential leads to a boost force on "boosted atoms" to accelerate the events. The boost potential is automatically constructed as a function of predefined CV via MD canonical ensemble sampling. The CV is a function of the positions of boosted atoms. To apply the ABMD method to the dislocation nucleation, a relative displacement of adjacent atomic plane along the slip direction can be considered as the CV.

In a conventional MD simulation at 300 K and 2.8 GPa, the $\Sigma 9$ GB emits partial dislocations using the collective multiple-dislocation nucleation mechanism on a timescale of picoseconds (Fig. 2.5a, b). However, at lower uniaxial tensile stress (similar to that at 300 K and 2.5 GPa, wherein only accelerated MD (i.e., ABMD) can be employed) a shuffling-assisted single-dislocation nucleation first occurs on a timescale of seconds (Fig. 2.5c, d). This finding suggests that the free-energy barrier exhibited by the shuffling-assisted single-dislocation nucleation mechanism is lower than that exhibited by the collective multiple-dislocation nucleation mechanism at lower stresses. To calculate the activation free energy $Q(\sigma, T)$ at finite temperatures, the nucleation frequency $\nu(\sigma, T)$ provided by ABMD or conventional MD is associated with $Q(\sigma, T)$ as

$$\nu(\sigma, T) = N\nu_0 \exp\left(-\frac{Q(\sigma, T)}{k_b T}\right) \tag{2.8}$$

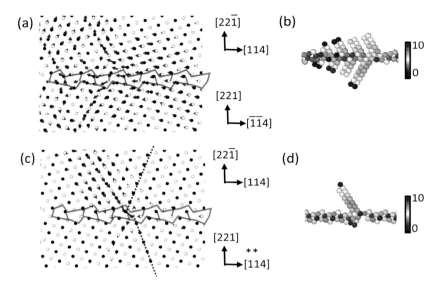

Fig. 2.5 **a** Atomistic displacement vector field of the collective multiple-dislocation nucleation mechanism and **b** the corresponding partial dislocation colored as per centrosymmetry (Kelchner et al. 1998). **c** Atomistic displacement vector field of the shuffling-assisted single-dislocation nucleation mechanism and **d** the corresponding partial dislocation colored as per centrosymmetry. Du et al. (2016) Reprinted figure with permission from [Du JP, Wang YJ, Lo YC, Wan L, Ogata S, Physical Review B, 94, 104110 (2016).] Copyright (2016) by the American Physical Society

where $nu_0 \sim 10^{11}\ s^{-1}$ denotes the attempt frequency, which is calculated using the curvature of the MEP (Zhu et al. 2008) N indicates the number of equivalent nucleation sites, and k_B denotes the Boltzmann constant. Figure 2.6a shows the activation free energies. There is a clear crossover (shoulder of each plot) of two mechanisms, such as the shuffling-assisted single-dislocation nucleation and collective multiple-dislocation nucleation, at all temperatures. Additionally, $Q(\sigma, T)$ dramatically decreases with an increase in T at specific σ, indicating a strong temperature dependence in the partial dislocation nucleation rate and large positive activation entropy. In experiments, deformation tests are performed at a constant strain rate in many cases, and the strain rate dependence of the dislocation nucleation stress is estimated. In nanocrystalline metals, the dislocation nucleation stress is directly related to the yield stress, because the dislocation activities in the grain are fairly restricted. At specific tensile strain rates, the critical dislocation nucleation stress can be obtained by solving the following equation (Zhu et al. 2008; Weinberger et al. 2012):

$$\frac{Q(\sigma, T)}{k_B T} = \ln \left[\frac{k_B T N \nu_0}{E \dot{\varepsilon} \Omega (\sigma, T)} \right] \tag{2.9}$$

where E denotes the apparent Young's modulus, $\dot{\varepsilon}$ represents the strain rate, ν_0 denotes the attempt frequency, and N represents the number of equivalent nucleation

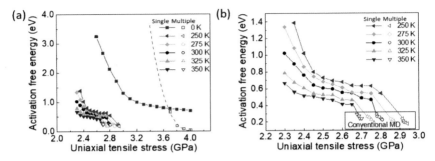

Fig. 2.6 **a** Activation free energies for shuffling-assisted single-dislocation nucleation (solid symbols) and collective multiple-dislocation nucleation (open symbols) mechanism. The free-end nudged elastic band method was used for the 0 K energy profile. ABMD (solid symbols) and conventional MD (open symbols) were used to construct the finite-temperature energy profiles. **b** Magnified view of (**a**) (Du et al. 2016). Reprinted figure with permission from [Du JP, Wang YJ, Lo YC, Wan L, Ogata S, Physical Review B, 94, 104110 (2016).] Copyright (2016) by the American Physical Society

sites. For simplification, activation energy $Q(\sigma, T)$ in Fig. 2.6 was fitted using analytical functions for the two mechanisms. The activation volume, i.e., the activation free-energy derivative with respect to stress can be calculated using the fitted analytical function. The mechanism with lower critical nucleation stress can be viewed as the dominant mechanism at a specific strain rate. Figure 2.7 presents the strain rate dependence of the critical nucleation stress. Here, the shuffling-assisted single-dislocation nucleation and collective multiple-dislocation mechanisms transpire at low strain rates (e.g., at 10^{-3} s^{-1} in the evaluated temperature range) and high strain rates (e.g., at 10^9 s^{-1} in the examined temperature range), respectively, which has been observed in the conventional high-strain rate MD simulation. The dislocation nucleation mechanism transition can be found as a kink of each plot, which cannot be detected in the conventional MD simulation. The mechanism transition can be observed even in actual experiments at experimentally feasible strain rates and temperatures, including at $\sim10^1 s^{-1} \sim10^3 s^{-1}$ and 300 K.

Hence, state-of-the-art atomistic modeling and free-energy-based analysis shed light on the possible mechanism transition of dislocation nucleation from GBs with respect to temperature and strain rates and its influences on mechanical properties.

2.4 Homogeneous Dislocation Nucleation in Nanoindentation (Sato et al. 2019)

In nanoindentation experiments, displacement bursts, known as "pop-in," are noticeable under load-controlled conditions. In particular, the first pop-in has been well studied because the indentation load at the first pop-in can be related to the ideal strength of the target material and thus the critical stress of homogeneous dislocation

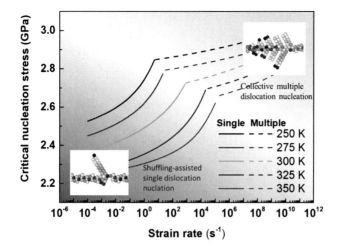

Fig. 2.7 Strain rate dependence of the critical nucleation stress for the shuffling-assisted single-dislocation nucleation and collective multiple-dislocation nucleation mechanisms (solid and dashed lines), respectively. A clear mechanism transition can be seen as a "kink" in each plot; this kink cannot be detected using the conventional MD simulation. Du et al. (2016) Reprinted figure with permission from [Du JP, Wang YJ, Lo YC, Wan L, Ogata S, Physical Review B, 94, 104110 (2016).] Copyright (2016) by the American Physical Society

nucleation (Shim et al. 2008; Morris et al. 2011; Li et al. 2012; Wu et al. 2015; Phani et al. 2013). Because the homogeneous dislocation nucleation is a thermally activated event, the first pop-in load exhibits strong temperature and loading rate dependencies, actually following the thermal activation theory (Mann and Pethica 1996, 1999; Biener et al. 2007; Rajulapati et al. 2010; Franke et al. 2015; Schuh and Lund 2004; Schuh et al. 2005; Mason et al. 2006). Recently, an atomistically informed prediction for the temperature and loading rate dependencies of the first pop-in load was achieved by formulating a homogeneous nucleation rate based on the free-energy analysis.

Sato et al. (2019) proposed an atomistic modeling-based multiscale (two-scale) method that avoids the timescale issue and consists of three steps. This method was verified for BCC Fe and Ta using the embedded atom method interatomic potentials (Mendelev et al. 2003; Ravelo et al. 2013). They first performed a simple MD simulation under various indentation loads below the first pop-in load and determined the stress state beneath the indenter (Step 1). Thereafter, they conducted NEB analysis (Henkelman and Jönsson 2000) of homogeneous dislocation nucleation using a perfect crystal atomic model by superimposing the stress state at each indentation load obtained in Step 1 to determine the indentation load-dependent activation energy for the homogeneous dislocation nucleation (Step 2). Next, using the indentation load-dependent activation energy obtained in Step 2, they analytically estimated the probability distribution and critical load of the pop-in event, as well as the corresponding

Fig. 2.8 Atomic model for Step 1 nanoindentation simulation of **a** Fe and **b** Ta (Sato et al. 2019)

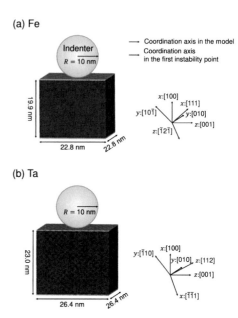

temperature and loading rate dependencies (Step 3). Further details regarding each step are described as follows.

(Step1) By the direct MD nanoindentation simulation to an atomistic model (Fig. 2.8), the local stress tensor was determined via the atomic stress tensor analysis (Thompson et al. 2009) with the Voronoi atomic volume (Du et al. 1999; Rycroft 2009). At the first instability points for Fe and Ta, the stress tensor components along the slip plane (in the coordinate system shown in Fig. 2.8) are

$$\sigma_{max}^{Fe} = \begin{bmatrix} -37.8 & -23.3 & 10.2 \\ -23.3 & -42.1 & 10.5 \\ 10.2 & 10.5 & -20.4 \end{bmatrix} \text{GPa,} \tag{2.10}$$

$$\sigma_{max}^{Ta} = \begin{bmatrix} -21.1 & 13.5 & 4.7 \\ 13.5 & -25.4 & -5.0 \\ 4.7 & -5.0 & -10.6 \end{bmatrix} \text{GPa.} \tag{2.11}$$

These stress states are very complicated, and are far from the pure shear condition. The non-shear components are known to change the critical stress of lattice instability owing to the elastic anisotropy and the non-linear elasticity of the materials. Interestingly, although they shear the same BCC structure, Fe and Ta exhibit different stress states. This difference is attributed to their different elastic anisotropies. Consequently, the first instability point (dislocation nucleation point) also differs between Fe and Ta (Fig. 2.9).

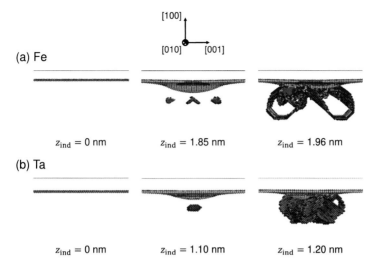

Fig. 2.9 Images indicating before and after the first pop-in of **a** Fe and **b** Ta (Sato et al. 2019); colors denote to the central symmetry parameter (Kelchner et al. 1998)

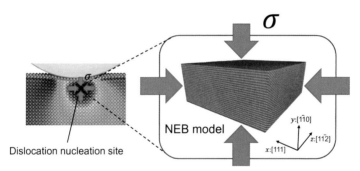

Fig. 2.10 Superimposition of the stress state beneath the indenter onto the NEB model (Sato et al. 2019)

(Step 2) The activation energy–indentation load relation of homogeneous dislocation nucleation, was determined using the NEB method. To mimic the dislocation nucleation event beneath the indenter under the actual indentation load using a perfect crystal model, the actual stress at the position exhibits the maximum resolved shear stress, as obtained in Step 1. This stress was superimposed onto a supercell that contains a perfect crystal (Fig. 2.10) by deforming the supercell shape. During NEB computation, the strained supercell shape was fixed. A perfect crystal without and with a dislocation loop on the $(1\bar{1}0)$ plane can be reasonably selected as the initial and final NEB images, respectively. Figure 2.11 presents the typical energy change along the minimum energy path and corresponding dislocation loop state (Fe and Ta at $P = 1.73 \times 10^{-2}$ µN and $P = 4.01 \times 10^{-2}$ µN and $z_{\text{ind}} = 0.080$ nm and $z_{\text{ind}} = 0.102$ nm, respectively). The indentation load-dependent activation

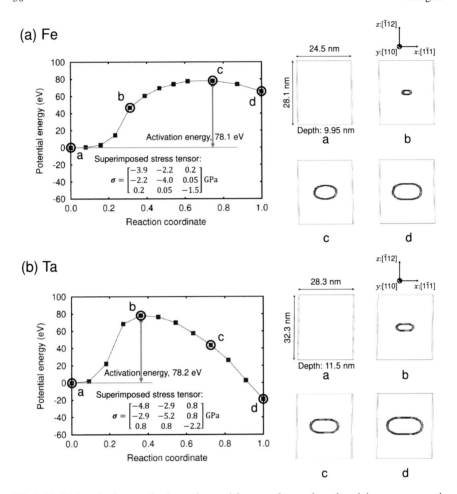

Fig. 2.11 Dislocation loop nucleation and potential energy change along the minimum energy path at the position showing the maximum resolved shear stress (Sato et al. 2019)

energy, $E(P)$, corresponds to the potential energy difference between the initial (perfect crystal) and saddle point (with an embryonic dislocation loop) configurations. Figure 2.12 presents $E(P)$, which is normalized by the square of the indenter radius R. $E(P)$ monotonically decreased with increasing indentation load (Barnoush et al. 2010) in both Fe and Ta. Ta exhibited a higher activation energy than Fe. $E(P)$, which can be reasonably fitted using the form proposed by Kocks et al. (1975), $E(P) = E_0\{1 - (P/P_c)^\alpha\}^\beta$, where E_0 denotes the activation energy under the stress-free condition, P_c represents the indentation load at the instability point, and α and β are parameters.

Such a two-scale modeling, including an MD indentation simulation to obtain the stress distribution (scale 1) and NEB analysis to determine the activation energy

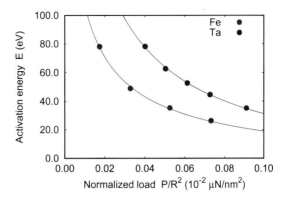

Fig. 2.12 Indentation load-dependent activation energy $E(P)$ for dislocation loop nucleation. The solid lines are the fitted curves (Sato et al. 2019)

(scale 2), implicitly assumes the scale invariance of the stress distribution beneath the indenter. The assumption is reasonable when the indenter radius is sufficiently larger than the dislocation loop size at the saddle point. In other words, the stress must be approximately constant within the dislocation loop area.

(Step3) Finally, to estimate the temperature and loading rate dependencies of the first pop-in load, the probability distribution of dislocation nucleation with respect to the indentation load must be determined. As per transition-state theory, probability distribution $p(P)$ and cumulative probability $Q(P)$ of the pop-in event can be expressed as

$$p(P) = \frac{k(P) \exp\left[-\dot{P}^{-1} \int_0^P k(P') dP'\right]}{\int_0^{P_c} k(P) \exp\left[-\dot{P}^{-1} \int_0^P k(P') dP'\right] dP}, \qquad (2.12)$$

$$Q(P) = \int_0^P p(P) dP, \qquad (2.13)$$

where

$$k(P) = \sum_i^N k_0(\mathbf{R}_i) \exp\left(-\frac{G(P, \mathbf{R}_i)}{k_B T}\right), \qquad (2.14)$$

$$G(P, \mathbf{R}_i) = E(P, \mathbf{R}_i) \left(1 - \frac{T}{T_m}\right). \qquad (2.15)$$

$k(P)$ denotes the load P-dependent dislocation nucleation rate and $G(P, \mathbf{R}_i)$ is the activation-free energy with temperature-dependent factor $(1 - T/T_m)$ (Zhu et al. 2008), where T indicates the absolute temperature, T_m can be approximately set to the melting temperature, k_B denotes the Boltzmann constant, k_0 denotes the attempt frequency at nucleation site i, $\mathbf{R}_i$ indicates the position of the possible nucleation site i, and N indicates the number of possible nucleation sites. $E(P, \mathbf{R}_i)$ is refers

Fig. 2.13 Predicted
temperature dependence of
the pop-in cumulative
probability for **a** Fe and **b**
Ta. The load is normalized
by the square of R_{sim} (Sato
et al. 2019)

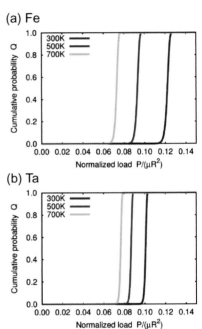

to the activation energy at 0 K at nucleation site i. Here, nucleation at the maximum resolved shear stress site is assumed to dominate the nucleation rate $k(P)$ because the high-stress spot should be well localized. Moreover, the dislocation nucleation rate exponentially decreases with shear stress; thus,

$$k(P) \approx N_{\text{eq}} k_0(\boldsymbol{R}_{\text{MRSS}}(P)) \exp\left(-\frac{G(P, \boldsymbol{R}_{\text{MRSS}}(P))}{k_{\text{B}}T}\right), \qquad (2.16)$$

where, $N_{\text{eq}} = 8$ for BCC, which equals the number of equivalent slip systems, and $\boldsymbol{R}_{\text{MRSS}}(P)$ denotes the position exhibiting the maximum resolved shear stress at indentation load P. Figures 2.13 and 2.14 present the calculated temperature and loading rate dependencies of cumulative probability $Q(P)$. The first pop-in load decreases with increasing temperature and decreasing loading rate, as observed in other experimental studies (Biener et al. 2007; Rajulapati et al. 2010; Franke et al. 2015) and expected based on theory (Schuh and Lund 2004). The pop-in load of Fe is more sensitive to both temperature and loading rate that that of Ta.

Hence, atomistically informed two-scale modeling and free-energy landscape analysis allow us to predict the pop-in load and its temperature and loading rate dependencies, which cannot be evaluated using the conventional MD simulation methods because of spatial and temporal scale limitations.

Fig. 2.14 Predicted
nanoindentation loading rate
dependence of pop-in
cumulative probability at
$T = 300$ K for **a** Fe and **b**
Ta. The load is normalized
by the square of R_{sim} (Sato
et al. 2019)

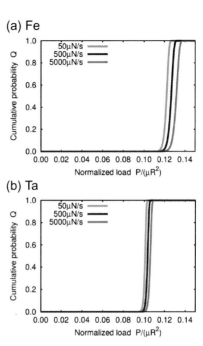

2.5   Summary

The three free-energy-based atomistic studies on defect nucleation were briefly introduced. The state-of-the-art atomistic modeling methods enable us to accurately elucidate the free-energy landscape of defect nucleation, as well as migration and propagation under different temperatures and stresses, resulting in a full description of the kinetics and thermodynamics of the defects. The next step is to determine how to control the kinetics and thermodynamics, i.e., how to design the free-energy landscape and tune materials' plasticity to manage their mechanical properties. Mechanical properties are well known to be tuned through the control of the loading conditions, such as temperature and strain rate, and possibly through the control of the chemical environment and the management of the texture and its evolution in materials and alloying. However, deducing the best conditions remains challenging because it is a highly nonlinear inverse problem. Combining atomistic free-energy analysis and machine learning technique can be a promising approach for overcoming this challenge.

Acknowledgements This work was supported by the Elements Strategy Initiative for Structural Materials (ESISM) of MEXT (Grant number JPMXP0112101000).

References

Barnoush A, Dake J, Kheradmand N, Vehoff H (2010) Examination of hydrogen embrittlement in FeAl by means of in situ electrochemical micropillar compression and nanoindentation techniques. Intermetallics 18:1385–1389. https://doi.org/10.1016/j.intermet.2010.01.001

Biener MM, Biener J, Hodge AM, Hamza AV (2007) Dislocation nucleation in bcc Ta single crystals studied by nanoindentation. Phys Rev B 76:165422–1–6. https://doi.org/10.1103/PhysRevB.76.165422

Christian JW, Mahajan S (1995) Deformation twining. Prog Mater Sci 39:1–157

Du Q, Faber V, Gunzburger M (1999) Centroidal Voronoi tessellations: applications and algorithms. SIAM Rev 41:637–676. https://doi.org/10.1137/S0036144599352836

Du JP, Wang YJ, Lo YC, Wan L, Ogata S (2016) Mechanism transition and strong temperature dependence of dislocation nucleation from grain boundaries: an accelerated molecular dynamics study. Phys Rev B 94(104):110. https://doi.org/10.1103/PhysRevB.94.10411

Franke O, Alcalá J, Dalmau R, Duan ZC, Biener J, Biener MM, Hodge AM (2015) Incipient plasticity of single-crystal tantalum as a function of temperature and orientation. Philos Mag 95:1866–1877. https://doi.org/10.1080/14786435.2014.949324

Henkelman G, Jönsson H (2000) Improved tangent estimate in the nudged elastic band method for finding minimum energy paths and saddle points. J Chem Phys 113:9978–9985. https://doi.org/10.1063/1.1323224

Ishii A, Ogata S, Kimizuka H, Li J (2012) Adaptive-boost molecular dynamics simulation of carbon diffusion in iron. Phys Rev B 85(064):303. https://doi.org/10.1103/PhysRevB.85.064303

Ishii A, Li J, Ogata S (2013) "Conjugate channeling" effect in dislocation core diffusion: carbon transport in dislocated bcc iron. PLoS ONE 8:e60586-1–7. https://doi.org/10.1371/journal.pone.0060586

Ishii A, Li J, Ogata S (2016) Shuffling-controlled versus strain-controlled deformation twinning: the case for HCP Mg twin nucleation. Int J Plas 82:32–43

Jonsson H, Mills G, Jacobsen KW (1998) Nudged elastic band method for finding minimum energy paths of transitions. In: Berne BJ, Ciccotti G, Coker DF (eds) Classical and quantum dynamics in condensed phase simulations. World Scientific, Singapore

Kelchner CL, Plimpton SJ, Hamilton JC (1998b) Dislocation nucleation and defect structure during surface indentation. Phys Rev B 58:11085–11088. https://doi.org/10.1103/PhysRevB.58.11085

Kocks UF, Argon AS, Ashby MF (1975) Thermodynamics and kinetics of slip, vol 19. Pergamon Press

Li B, Ma E (2009) Atomic shuffling dominated mechanism for deformation twinning in magnesium. Phys Rev Lett 103: 035503-1-4. https://doi.org/10.1103/PhysRevLett.103.035503

Li TL, Bei H, Morris JR, George EP, Gao YF (2012) Scale effects in convoluted thermal/spatial statistics of plasticity initiation in small stressed volumes during nanoindentation. Mater Sci Technol 28:1055–1059. https://doi.org/10.1179/1743284712Y.0000000007

Mann AB, Pethica JB (1996) The role of atomic size asperities in the mechanical deformation of nanocontacts. Appl Phys Lett 69:907–909. https://doi.org/10.1063/1.116939

Mann AB, Pethica JB (1999) The effect of tip momentum on the contact stiffness and yielding during nanoindentation testing. Philos Mag A Phys Cond Matter Struct Defects Mech Prop 79:577–592. https://doi.org/10.1080/01418619908210318

Mason JK, Lund AC, Schuh CA (2006) Determining the activation energy and volume for the onset of plasticity during nanoindentation. Phys Rev B 73:054102–1–14. https://doi.org/10.1103/PhysRevB.73.054102

Mendelev MI, Han S, Srolovitz DJ, Ackland GJ, Sun DY, Asta M (2003) Development of new interatomic potentials appropriate for crystalline and liquid iron. Phil Mag 83:3977–3994. https://doi.org/10.1080/14786430310001613264

Morris JR, Bei H, Pharr GM, George EP (2011) Size effects and stochastic behavior of nanoindentation pop in. Phys Rev Lett 106:165502–1–4. https://doi.org/10.1103/PhysRevLett.106.165502

Ogata S, Li J, Yip S (2002) Ideal pure shear strength of aluminum and copper. Science 298:807–811. https://doi.org/10.1126/science.1076652

Ogata S, Li J, Yip S (2005) Energy landscape of deformation twinning in bcc and fcc metals. Phys Rev B 71:224102-1-11. https://doi.org/10.1103/PhysRevB.71.224102

Phani PS, Johanns KE, George EP, Pharr GM (2013) A stochastic model for the size dependence of spherical indentation pop-in. J Mater Res 28:2728–2739. https://doi.org/10.1557/jmr.2013.254

Rajulapati KV, Biener MM, Biener J, Hodge AM (2010) Temperature dependence of the plastic flow behavior of tantalum. Philos Mag Lett 90:35–42. https://doi.org/10.1080/09500830903356893

Ravelo R, Germann TC, Guerrero O, An Q, Holian BL (2013) Shock-induced plasticity in tantalum single crystals: interatomic potentials and large-scale molecular-dynamics simulations. Phys Rev B 88:134101–1–17. https://doi.org/10.1103/PhysRevB.88.134101

Rycroft CH (2009) Voro++: a three-dimensional Voronoi cell library in C++. Chaos Interdiscipl J Nonlinear Sci 19:041111. https://doi.org/10.1063/1.3215722

Sato Y, Shinzato S, Ohmura T, Ogata S (2019) Atomistic prediction of the temperature- and loading-rate dependent first pop-in load in nanoindentation. Int J Plast 121:280–292. https://doi.org/10.1016/j.ijplas.2019.06.012

Schuh CA, Lund AC (2004) Application of nucleation theory to the rate dependence of incipient plasticity during nanoindentation. J Mater Res 19:2152–2158. https://doi.org/10.1557/JMR.2004.0276

Schuh CA, Mason JK, Lund AC (2005) Quantitative insight into dislocation nucleation from high-temperature nanoindentation experiments. Nat Mater 4:617–621. https://doi.org/10.1038/nmat1429

Sheppard D, Xiao P, Chemelewski W, Johnson DD, Henkelman G (2012) A generalized solid-state nudged elastic band method. J Chem Phys 136(074):103

Shim S, Bei H, George EP, Pharr GM (2008) A different type of indentation size effect. Scripta Mater 59:1095–1098. https://doi.org/10.1016/j.scriptamat.2008.07.026

Thompson AP, Plimpton SJ, Mattson W (2009) General formulation of pressure and stress tensor for arbitrary many-body interaction potentials under periodic boundary conditions. J Chem Phys 131:154107–1–6. https://doi.org/10.1063/1.3245303

Wang J, Li J, Yip S, Phillpot S, Wolf D (1995) Mechanical instabilities of homogeneous crystals. Phys Rev B 52:12627–12635. https://doi.org/10.1103/PhysRevB.52.12627

Wang J, Yadav SK, Hirth JP, Tomé CN, Beyerlein IJ (2013) Pure-shuffle nucleation of deformation twins in hexagonal-close-packed metals. Mater Res Lett 1:126–132

Wang YJ, Ishii A, Ogata S (2011) Transition of creep mechanism in nanocrystalline metals. Phys Rev B 224102. https://doi.org/10.1103/PhysRevB.84.22410

Weinberger C, Jennings A, Kang K, Greer J (2012) Atomistic simulations and continuum modeling of dislocation nucleation and strength in gold nanowires. J Mech Phys Solids 60:84–103. https://doi.org/10.1016/j.jmps.2011.09.010

Wu D, Morris JR, Nieh TG (2015) Effect of tip radius on the incipient plasticity of chromium studied by nanoindentation. Scripta Mater 94:52–55. https://doi.org/10.1016/j.scriptamat.2014.09.017

Zhu T, J Li AS, Leach A, Gall K (2008) Temperature and strain-rate dependence of surface dislocation nucleation. Phys Rev Lett 100:025502–1–4. https://doi.org/10.1103/PhysRevLett.100.025502

Chapter 3
Atomistic Study of Disclinations in Nanostructured Metals

Tomotsugu Shimokawa

3.1 Introduction

3.1.1 Various Deformation Modes in Nanostructured Metals

Nanostructured materials generally have unique mechanical properties that cannot be easily expressed by extrapolating the mechanical properties of coarse-grained materials (Gleiter 2000; Valiev et al. 2000; Meyers et al. 2006; Yinmin et al. 2002; Lu et al. 2009; Wang et al. 2011). A possible reason for these unique mechanical properties is that the activated plastic deformation modes in nanostructured materials are different from those in coarse-grained materials. Intrinsically, each crystalline material has various plastic deformation modes: *plaston* (Tsuji et al. 2020), as discussed in detail in the previous chapters of this book. Among the several candidates of plastic deformation modes, one is strongly activated in relation to factors such as constituent elements, structure size, and environment. For example, the number of dislocations in each structure decreases with the decrease in the structure size below the sub-micron level when the dislocation density does not depend on the structure size. For example, if the dislocation density is 10^{14} 1/m^2, there would be 10,000 dislocations in a 10 μm grain but only 1 dislocation in a 100 nm grain. Therefore, in nanostructured metals, the generation of a new plastic deformation mode from the interface between neighboring structures is more important than the plastic phenomena that start from the intragranular region. Then, unique lattice defects can develop in nanostructured materials.

An example of the structure dependence of activated deformation modes is shown in Fig. 3.1, which also shows the grain size dependence of the (a) (b) strength and (c) (d) deformation mechanism in nanopolycrystalline Al, which was obtained using

T. Shimokawa (✉)
Faculty of Mechanical Engineering, Kanazawa University, Kakuma-machi, Kanazawa, Ishikawa 920-1192, Japan
e-mail: simokawa@se.kanazawa-u.ac.jp

© The Author(s) 2022
I. Tanaka et al. (eds.), *The Plaston Concept*,
https://doi.org/10.1007/978-981-16-7715-1_3

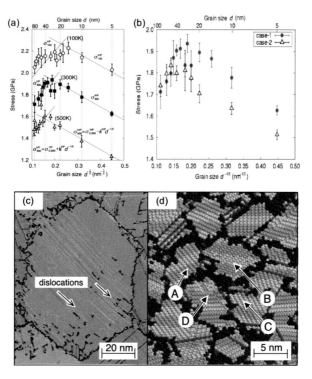

Fig. 3.1 Grain size dependence of the strength and deformation mechanism in nanostructured Al. **a** Temperature dependence of strength (Shimokawa 2012). **b** Grain boundary structure dependence of strength (Shimokawa 2012). **c** Collective dislocation motion constrained by the grain boundary ($d = 80$ nm) (Shimokawa et al. 2005). **d** Grain rotation of grains A and B and grain boundary sliding between grains C and D ($d = 5$ nm) (Shimokawa et al. 2005). Reprinted with permission from Springer and American Physical Society

molecular dynamic simulations (Shimokawa et al. 2005; Shimokawa 2012). The transition from the grain size strengthening (Hall-Petch relation) to grain size softening (inverse Hall-Petch relation) (Chokshi et al. 1989; Fougere et al. 1992; Schiøtz and Jacobsen 2003) occurred as shown in Fig. 3.1a, b. In the Hall-Petch region, dislocation pile-ups against the grain boundary can be observed in Fig. 3.1c. Hence, the dominant deformation mechanism was caused by intragranular deformation modes. However, in the inverse Hall-Petch region, the geometrical misfit caused by the grain boundary sliding of grains C and D can be accommodated by the grain rotation of grains A and B, as shown in Fig. 3.1d. Thus, the dominant deformation mechanism was caused by intragranular deformation modes. Figure 3.1a shows the temperature dependence of the strength and grain size relation and the maximum grain size changes with the following temperatures: 20 nm at 100 K, 30 nm at 300 K, and 40 nm at 500 K (Shimokawa 2012). The black lines represent the fitting results of the Hall-Petch relation: $\sigma = \sigma_0 + k d^{-1/2}$. Also, k does not show the temperature dependence but σ_0 shows it. This is because the grain boundary-mediated plastic deformation

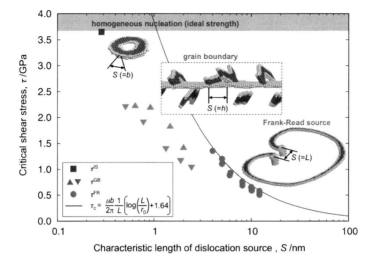

Fig. 3.2 Relationship between the critical shear stress and the characteristic length of the dislocation source of Al estimated by molecular dynamics simulations. Reproduced from Shimokawa and Kitada (2014) with permission from The Japan Institute of Metals and Materials

contains more thermal-activated processes than intragranular deformation. On the other hand, Fig. 3.1b shows the grain boundary structure dependence of the strength and grain size relation (Shimokawa 2012). Case-1 and case-2 have the same texture but different grain boundary structures by changing the grain positions. In the strengthening region, case-1 is stronger than case-2, but in the softening region, case-2 is stronger than case-1. These results show that the role of the grain boundary in strength is influenced by the structure. Consequently, the active deformation modes were changed by the grain size and temperature, and this phenomenon influences the mechanical properties of nanostructured metals.

Figure 3.2 shows the relationship between the critical shear stress to nucleate the dislocation and the characteristic length (Shimokawa and Kitada 2014). The solid circles, triangles, and squares represent the dislocation multiplication results from a Frank-Read source, the dislocation emission from the ⟨112⟩ tilt grain boundaries, and the homogeneous nucleation from a perfect crystal, respectively (Shimokawa and Kitada 2014). The characteristic length of the grain boundary is the distance between the structural units that can act as a dislocation source (Shimokawa 2010), and the homogeneous nucleation length is the Burgers vector. This result clearly shows that the active deformation mode changes with the decrease in the characteristic length.

As a result, the deformation mode *plaston* can be selected by designing structures, elements, and environments. Afterward, materials with excellent mechanical properties can be obtained. Recent attempts have been made to design materials with excellent mechanical properties by making heterogeneous microstructures, such as gradient nanotwinned metals (Cheng et al. 2018; Sun et al. 2018) and harmonic materials (Zhang et al. 2014; Sawangrat et al. 2014). Such excellent properties might be closely related to a mix of various structure-specific deformation modes in such heterogeneous structure materials.

3.1.2 Disclinations

In this chapter, we considered disclination as an example of plaston. Disclination is a line defect in which rotational symmetry is violated (Romanov and Vladimirov 1992). Although it is generally difficult for crystalline materials to contain disclinations in a stable form in intragranular regions, disclinations have recently been recognized as typical defects that can influence the mechanical properties of nanostructured metals (Valiev et al. 2000; Gutkin and Ovid'ko 2004), as seen in nanostructure materials (Murayama et al. 2002). Moreover, the stability of disclinations at the grain boundaries of bicrystalline nanowires has been studied using computer simulations (Zhou et al. 2006, 2007). Here, we introduced three examples of disclinations that can be observed in nanostructured materials under deformation through atomic simulations.

The first example is disclinations in a grain, as shown in Fig. 3.3a. Since a perfect disclination is hard to exist in grains due to the huge elastic strain energy, generally partial disclinations with plane defects exist in grains. Plane defects have a misorientation angle; hence, partial disclinations can divide a grain into several regions with misorientation angles. In Sect. 3.2, we discussed the grain refinement mechanism by severe plastic deformation focusing on the formation and movement of partial disclinations (Shimokawa et al. 2016). The next example shows disclinations at the junctions of grain boundaries, as shown in Fig. 3.3b. The formation of these disclinations brings grain rotation (Shimokawa et al. 2005, 2006), as already shown in Fig. 3.1d. The possible mechanism of the grain rotation with disclinations was proposed by Gutkin et al. (2003). The last example shows disclinations at the grain boundaries, as shown in Fig. 3.3c. The disclination dipole at a grain boundary changes the misorientation angle at the grain boundary region, and its stress field is formed in the grain. These stress fields can improve the fracture toughness of nanostructured

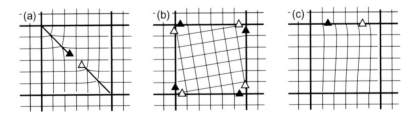

Fig. 3.3 Disclinations **a** in a grain, **b** at the grain boundary junctions, and **c** at the grain boundary. Each disclination state brings grain subdivision (Shimokawa et al. 2016), grain rotation (Shimokawa et al. 2005), and improved fracture toughness (Shimokawa et al. 2011; Shimokawa and Tsuboi 2015), respectively. The thick and thin solid lines represent the grain boundary and a common lattice of grains, respectively

metals (Shimokawa et al. 2011; Shimokawa and Tsuboi 2015). A detailed discussion can be found in Sect. 3.3.

3.2 Grain Subdivision: Disclinations in Grains

3.2.1 Strain Gradients in Severe Plastic Deformation Processes

The severe plastic deformation (SPD) process is a method of producing nanostructured metals in bulk. As shown in Fig. 3.4, the main types of SPD are equal-channel angular pressing (ECAP) (Iwahashi et al. 1996; Valiev and Langdon 2006), accumulative roll-bonding (Saito et al. 1999), and high-pressure torsion (HPT) (Valiev et al. 1996). Bulk ultra-fine-grained (UFG) metals, which could not be obtained before, could be produced by repeatedly introducing large strains into specimens. Hence, the research on the mechanical properties of UFG metals has dramatically developed. The common feature of these SPD processes is the introduced strain gradient into specimens. In the cases of ECAP and ARB, the strain gradient appeared at the border between the deformed and undeformed regions, and in the case of HPT, the strain gradient appeared along the radius direction. In the strain gradient regions, there were dislocations that could accommodate the plastic deformation gaps. These dislocations are called geometrically necessary dislocations (GNDs) (Ashby 1970). As a result, a large number of GNDs are introduced during SPD processes, and this phenomenon causes grain refinement. The concept of *grain subdivision* was proposed by Hansen et al. (2001) to explain grain refinement through SPD, where an original grain domain is finely divided by rearranging several GNDs, which are introduced by SPD into energetically stable structures.

However, it is still not clear why introducing GNDs into specimens results in grain subdivisions due to the difficulty of directly observing the structural changes in bulky metallic specimens during SPD processes. Hence, molecular dynamic simulations of the SPD process were performed using the analysis model shown in Fig. 3.5 (Shimokawa et al. 2016). To simplify the SPD process, a quasi-two-dimensional model was employed. The crystal structure is a hexagonal close-packed structure, where two {0001} planes are arranged along the y-direction. Adopting the periodic boundary condition in the y-direction realized three equivalent prismatic slip systems. The outer region of the analysis model was set to the displacement controlled layer, and the shear strain $\gamma = \lambda/A$ was introduced by changing the shape of the layer along the z-direction at a velocity v. The strain gradient appeared at the border of the shear and non-shear zones. Our SPD process is similar to the multi-pass ECAP (Nakashima et al. 2000) and ECAP in parallel channels (Raab 2005). Two types of crystal orientations, as shown in Fig. 3.5c, were adopted in the analysis models to investigate the crystal orientation influence on the activated slip systems in the vicinity of the strain gradient regions. The embedded atom method for copper (Mishin

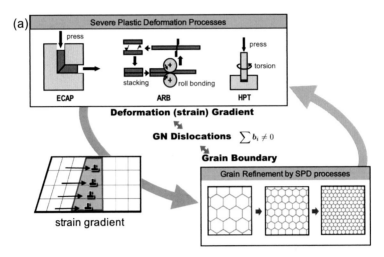

Fig. 3.4 Relationship among the severe plastic deformation processes, strain gradient, and grain refinement

Fig. 3.5 a, b Analysis model for investigating the relationship through molecular dynamic simulations. **c** Two types of crystal orientation were used in the analysis models. Reproduced from Shimokawa et al. (2016) with permission from The Japan Institute of Metals and Materials

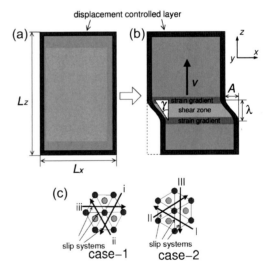

et al. 2001) was used for the atomic interactions. Although the most stable phase of the EAM potential is the face-centered cubic phase, the transformation from the hexagonal close-packed phase to the face-centered cubic phase did not occur during the SPD simulations because of the boundary conditions.

Figure 3.6 shows the crystal orientation influence on the activated slip systems around the strain gradient regions. The crystal orientation of case-1 has the slip system iii parallel to the x-direction of the shear deformation direction, and the crystal orientation of case-2 has the slip system III orthogonal to the slip system iii. The analysis conditions are $\gamma = 0.2$, $v = 500$ m/s, and $T = 300$ K. Figure 3.6a, b

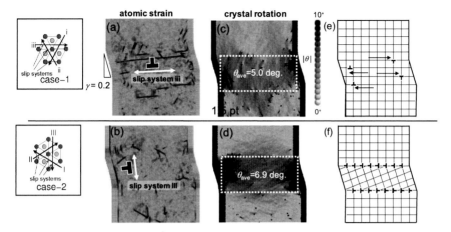

Fig. 3.6 Influence of the crystal orientation on the activated slip systems at the vicinity of the strain gradient regions with $\gamma = 0.2$. **a, c** Crystal slip histories colored by the atomic strain. **b, d** Crystal rotation angle from the initial state

shows the atomic strain during shear deformation, where the black lines represent the crystal slip histories. In case-1, the slip system iii was mainly activated. However, in case-2, the primally slip system was III. As mentioned before, these slip systems are orthogonal to each other, so the same shear stress magnitude was applied to all of them. Figure 3.6c, d shows the change in the crystal orientation for each crystal orientation. The atomic color represents the crystal rotation angle from the initial state. The average rotation angle θ_{AVE} in the shear zone in case-2 was larger than that in case-1. Hence, the accommodation mechanism of the strain gradient changed due to the activated slip systems, as shown in Fig. 3.6e, f. In this chapter, SPD simulations were performed using case-2.

3.2.2 Grain Subdivision by Severe Plastic Deformation

Figure 3.7 shows the taken snapshots during the SPD simulations with $\gamma = 0.7$ and $v = 500$ m/s. The atomic color represents the crystal rotation angle from the initial state, and the black atoms are those in a defect structure. A clear difference in the crystal orientation appeared around the strain gradient regions, and the boundaries with a misorientation angle propagated along with the specimen. Finally, distinct regions with different crystal orientations were formed in the specimen after the SPD process, as shown in Fig. 3.7g, where grain subdivisions occurred in the simple atomic simulations.

Figure 3.8 shows the influence of the local shear strain γ on the microstructure after the SDP process. In the cases of small γ, as shown in Fig. 3.8a, b, no microstructures appeared after the SPD process. However, in the cases of larger γ, as shown in

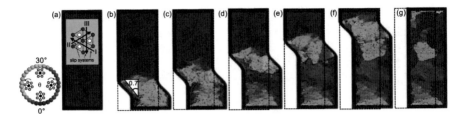

Fig. 3.7 Grain subdivision caused by the SPD process with the propagation of the local shear strain $\gamma = 0.7$. The atomic colors other than black represent the rotation angle from the initial state, and the black atoms represent the defect structures defined using the common neighbor analysis. Reproduced from Shimokawa et al. (2016) with permission from The Japan Institute of Metals and Materials

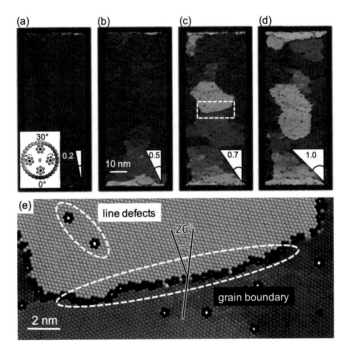

Fig. 3.8 Dependence of the local shear strain γ on the observed microstructures after the SPD process. Crystal orientation maps after the SPD processes of **a** $\gamma = 0.2$, **b** $\gamma = 0.5$, **c** $\gamma = 0.7$, and **d** $\gamma = 1.0$. **e** Atomic configurations around the grain boundary indicated by a white broken box in (**c**). Reproduced from Shimokawa et al. (2016) with permission from The Japan Institute of Metals and Materials

Fig. 3.11c, d, new grains were formed by the SPD process. Figure 3.8e shows the detailed atomic configurations in the broken box in Fig. 3.8c. Two types of defects can be observed: one is the line defects that do not bring crystal misorientations, and the other is the grain boundaries with a misorientation angle. To quantify the influence of γ on the microstructure formation, Fig. 3.9 shows the proportion of the defect atoms

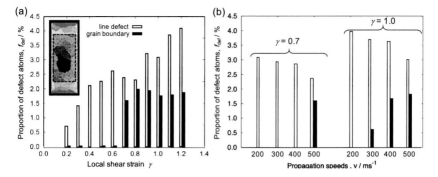

Fig. 3.9 Proportions of the defect atoms: line defect and grain boundary. The proportion of the defect atoms was calculated in the central region of the specimens, as shown in the inset of (**a**). **a** Local shear strain dependence ($v = 500$ m/s). **b** Propagation speed dependence ($\gamma = 0.7$ and 1.0). Reproduced from Shimokawa et al. (2016) with permission from The Japan Institute of Metals and Materials

f_{def} with respect to (a) γ and (b) v after the SPD process. Here, f_{def} was measured for each defect type, and the analysis region for f_{def} is the central region surrounded by the broken line shown in the insert figure in Fig. 3.9a. It was confirmed that f_{def} of the line defects increases with γ. Since most line defects are vacancy arrays formed by interactions between dislocations with the same slip system but opposite Burgers vectors, the f_{def} of the line defects reflected the number of dislocations activated during the SPD process. The results show that many dislocations were nucleated with the increase in γ. However, f_{def} of the grain boundaries showed almost zero for $\gamma \leq 0.6$ but showed a constant value for $\gamma \geq 0.7$. This indicates that a critical value of γ existed to realize the grain subdivision in the atomic simulations. Figure 3.8b shows the influence of the propagation speed of the local shear zone v on f_{def} for $\gamma = 0.7$ and 1.0. f_{def} of the grain boundaries was also influenced by v, as a smaller v could not form new microstructures in the specimen that was even larger than γ. These results imply that the necessary conditions for grain subdivision in SPD simulations not only contain geometrical aspects but also kinematic aspects.

3.2.3 Partial Disclinations Induced by the Strain Gradient

Figure 3.10a, b shows the atomic configurations at the vicinity of the grain gradient regions for $\gamma = 0.2$ and 0.7. The atomic color represents the rotation angle from the initial state. As shown in the figure, defect boundaries with misorientation angles appeared around the strain gradient regions. Figure 3.10d, e shows the defect distributions of the same areas of Fig. 3.10a, b. A defect analysis was performed by drawing the Burgers circuits for each defect atom, as shown in Fig. 3.10f. In both cases of $\gamma = 0.2$ and 0.7, geometrically necessary boundaries (GNBs) were formed

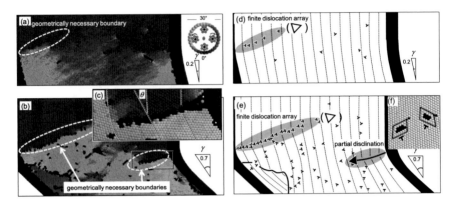

Fig. 3.10 Distributions of the lattice defects around the strain gradient regions during SPD processes. **a, c** $\gamma = 0.2$. **b, d** $\gamma = 0.7$. **c** A higher resolution image of the region marked out by a box in (**b**). **f** Burgers circuits for determining the defect structure types. Partial disclinations accompanied by finite dislocation arrays and a plane defect. The structures of the partial disclination boundaries change according to the local shear strain γ. The filled and unfilled triangles denote the positive and negative partial disclinations, respectively. Reproduced from Shimokawa et al. (2016) with permission from The Japan Institute of Metals and Materials

by the finite edge dislocation arrays with the same sign of the slip system for the left side of the strain gradient regions. The GNBs could accommodate the strain gradient fields. The distance between the dislocations in the GNBs became shorter with the increase in γ, as the misorientation angle at the GNBs became larger. With the increase in the misorientation angle, it is difficult to express GNBs as dislocation arrays. Hence, the GNB was expressed as a plane defect as shown on the right side of the strain gradient region in Fig. 3.10e. This plane defect did not cross the specimen, so a rotational type of defect field appeared in the specimen. As a result, this defect can be regarded as a partial disclination (Romanov and Vladimirov 1992) with a plane defect. Notice that the finite dislocation arrays on the left side also can be regarded as partial disclinations. That is, the plane defect structure transitions with the magnitude of the Frank vector ω, which represents the disclination strength, as shown in Fig. 3.11.

Partial disclination mobility is important for grain subdivisions, as strain gradient regions travel through specimens during the SPD process. If the partial disclination mobility was small, the GNBs would not have been able to follow the moving strain gradient regions in the specimen. Consequently, they would have remained in the specimen and formed a microstructure. Since partial disclinations are accompanied by grain boundaries, the mobility of grain boundaries is closely related to the grain subdivision process. Furthermore, the interactions between dislocations and partial disclinations are also important since many dislocations are generated in SPD processes. Also, the mobility of the partial disclinations interacting with lattice dislocations strongly depends on their grain boundary structure. When a structure is represented as a lattice dislocation array in the case of a small ω, as shown on the left

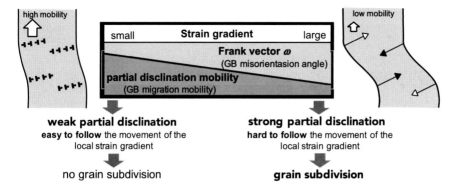

Fig. 3.11 Grain subdivision mechanism obtained by a severe plastic deformation process through molecular dynamic simulations

side of Fig. 3.11, it is hardly affected by external dislocations. However, when it is represented as a structural unit in the case of a large ω, as shown on the right side of Fig. 3.11, it is strongly affected by external dislocations, resulting in the formation of new microstructures by grain subdivision with GNBs remaining in the specimen (Shimokawa et al. 2016). The critical misorientation angle of the grain boundary transition in the partial disclination from the dislocation array to the structural units corresponded to the critical γ required for the grain subdivision in this simulation (Shimokawa et al. 2016). These are the geometrical and kinematic reasons for the existence of the critical local shear strain γ in the grain subdivision of the present simulation.

3.3 Fracture Toughness: Disclinations at the Grain Boundary

3.3.1 High Strength and High Toughness

In general, it is difficult to achieve both high strength and high fracture toughness with single-crystal materials from the viewpoint of dislocation mobility, but it has been reported that UFG materials with enhanced strength can achieve both high strength and high fracture toughness by reducing the grain size to the sub-micron order (Hodge et al. 1949; Tsuji et al. 2004; Tanaka et al. 2008). Since UFG materials have more grain boundaries than coarse-grained materials, it is important to understand the role of grain boundaries in crack-initiated fracture phenomena. In this section, a new shielding process was presented via grain boundaries (Shimokawa et al. 2011; Shimokawa and Tsuboi 2015) based on the lattice defects evolution, as shown in Figs. 3.12 and 3.13, and it may be one of the mechanisms for improving the fracture toughness of UFG materials based on the dislocation shielding theory (Rice and Thomson 1974; Majumdar and Burns 1981) and molecular dynamic simulations.

3.3.1.1 Dislocation Shielding

In general, there is a trade-off relationship between the yield stress and fracture toughness of crystalline materials with a high Peierls stress, and it is known that the fracture toughness decreases with the increase in the yield stress. Since the yield stress is negatively correlated with temperature, a brittle-ductile transition behavior can be observed for such materials at a certain temperature. Previous experiments have shown that the activation energy of the brittle-ductile transition is well correlated with the activation energy of dislocation mobility (Giannattasio et al. 2007), and it has been reported that the elemental mechanism controlling the behavior of brittle-ductile transition is the mobility of dislocations. In other words, as the dislocation motion is suppressed at lower temperatures, the material becomes more brittle, and the fracture toughness value decreases. This phenomenon can be explained by considering the dislocations shielding effect on the crack tip mechanical field, as shown in Fig. 3.12a (Rice and Thomson 1974). For a coarse-grained material in which the distance between the crack and grain boundary is sufficiently large, the mechanical field at the crack tip is shielded by the mechanical field of the dislocations emitted from the crack. This effect is called *dislocation shielding*. If the dislocation stays near the crack tip, it is difficult to emit subsequent dislocations from the crack tip, which then reaches a cleavage failure environment with a small stress intensity factor. Hence, to obtain a large fracture toughness value, the emitted dislocations from a crack tip have to move far enough away to shield it by a large number of dislocations. In other words, the mechanical field at a crack tip, which governs the fracture toughness, is determined by the competition between "the expansion speed of the stress concentrated field (K-field) with the increase in the external force" and "the expansion speed of the plastic region (dislocation mobility)".

3.3.1.2 Grain Boundary Shielding

As known in the Hall-Petch relation, grain refinement is one of the methods for strengthening materials. The researches on the fracture toughness of UFG materials have shown that at low temperatures, the brittle-ductile transition temperature is shifted to lower temperatures and that the fracture toughness is improved (Tsuji et al. 2004; Tanaka et al. 2008). In short, grain size refinement is expected to be an excellent method for achieving both high strength and high toughness. However, since the activation energies of the brittle-ductile transition between coarse- and fine-grained materials are almost the same (Tanaka et al. 2009), it can be deduced that the dislocation mobility is not affected by the grain size. Therefore, it is clear that to understand the grain size dependence of fracture toughness, it is necessary to extend the conventional shielding theory to consider another dominant factor that shows the

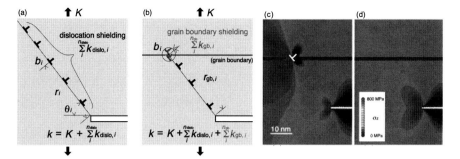

Fig. 3.12 Schematics of **a** the dislocation shielding and **b** grain boundary shielding. **c, d** Stress fields around the crack tip with/without a dislocation at the grain boundary under the same applied loading. A grain boundary shielding can be clearly confirmed in (**c**). Reprinted with permission from Shimokawa et al. (2011). Copyright 2011 by American Physical Society

grain size dependence in addition to the dominant factor of the dislocation mobility that controls the brittle-ductile transition behavior. The major difference between coarse and fine-grained materials is the volume fraction of grain boundaries. As shown in Fig. 3.12b, in the conventional shielding theory, if the grain boundaries role is considered as an obstacle to dislocations, a dislocation emitted from a crack can only move up to the grain boundaries. In this case, the shielding effect at the crack tip by the dislocations entering the grain boundary is termed as *grain boundary shielding*, which is clearly observed in Fig. 3.12c, d. The stress field ahead of the crack tip decreased with the existence of a dislocation at the grain boundary. With the decrease in the grain size, dislocations piled up against the grain boundary in the vicinity of the crack, and the back stress suppressed the release of dislocations from the crack tip. As a result, it is not possible to easily increase the number of dislocations emitted from a crack tip, and the fracture toughness is reduced (Noronha and Farkas 2004; Zeng and Hartmaier 2010). This result cannot explain the actual experimental results, so it was necessary to extend the role of grain boundaries to resolve this discrepancy.

3.3.1.3   Disclination Shielding

The atomic simulations of the interactions among dislocations, crack tip, and grain boundary showed that this contradiction can be resolved by considering the role of the grain boundary as a dislocation generation site (Shimokawa et al. 2011). If dislocations are continuously emitted from a grain boundary, disclination dipoles can be formed, as shown in Fig. 3.13. The stress field in this disclination dipole can reduce the stress field at the crack tip, meaning that the mechanical field at the crack tip can be shielded without losing the plastic deformation ability near the crack by the dislocations emission from the grain boundary. This is called *disclination shielding*. The details of the disclination shielding mechanism are described below.

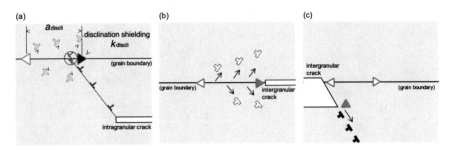

Fig. 3.13 Schematics of the disclination shielding with a dislocation emission from the grain boundary observed in atomic simulations to realize the improving fracture toughness. **a** Intragranular crack and **b**, **c** intergranular cracks. Reprinted with permission from Shimokawa et al. (2011). Copyright 2011 by American Physical Society

3.3.2 Dislocation Emission from the Grain Boundary

The grain boundary that achieved disclination shielding changed the grain boundary misorientation in the dislocation-emitted region. In other words, the grain boundary structure must transition to another grain boundary structure with the release of a dislocation. One grain boundary that satisfies this is the $\langle 112 \rangle$ symmetrical tilt grain boundary in the face-centered cubic system. Figure 3.14a shows the relationship between the grain boundary misorientation angle and the grain boundary energy of the Al $\langle 112 \rangle$ symmetrical tilt grain boundaries (Shimokawa 2010). The numbers near the plot represent the Σ value of each grain boundary. There is a $\Sigma 11$ grain boundary with minimum energy at $\theta = 62.96°$. Figure 3.14b, c shows the grain boundary structure of the $\Sigma 11$ grain boundary and the $\Sigma 15$ grain boundary, which has a larger misorientation angle than the $\Sigma 11$ grain boundary. The structure of the $\Sigma 11$ grain boundary can only be represented by the B-structural unit, while the structure of the $\Sigma 15$ grain boundary can be represented by the periodic arrangement of the B and C structural units. To investigate the relationship between the two structures, the displacement shift complete (DSC) lattice (Bollmann 1970) and coincidence site lattice (CSL) of the $\Sigma 11$ grain boundary were demonstrated as thin and thick solid lines, respectively, as shown in Fig. 3.14d. Figure 3.14e shows the DSC lattice of the $\Sigma 11$ grain boundary applied to the $\Sigma 15$ grain boundary. This indicates that there was a grain boundary dislocation in the C-structural unit of the $\Sigma 15$ grain boundary that formed a change in the misorientation angle from the $\Sigma 11$ grain boundary structure. The Burgers vector of the grain boundary dislocation is $b^{gb+} = \frac{2}{11}[1\bar{3}\bar{1}]_{II}$. Here, the Burgers vector of the slip system, which was activated by applying a tensile load perpendicular to the interface of the $\Sigma 15$ grain boundary, was $b^{lt} = \frac{1}{2}[1\bar{1}0]_{II}$, and the response of the Burgers vectors when a lattice dislocation was emitted using a grain boundary dislocation is as follows:

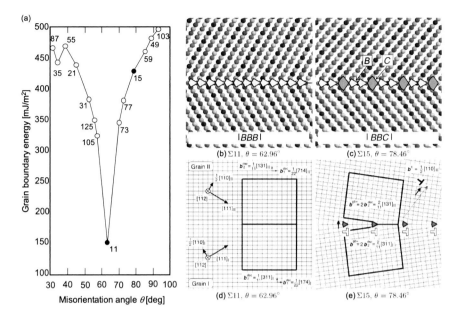

Fig. 3.14 a Grain boundary energy and misorientation angle relationship of the aluminum $\langle 112 \rangle$ tilt grain boundaries. The Σ values are shown near the symbols. **b, c** The grain boundary structures of $\Sigma 11$ and $\Sigma 15$ are represented by the B and C structural units. **d, e** The DSC lattices of $\Sigma 11$ and $\Sigma 15$. The atomic colors represent the different depths of the stacking atomic layers along the $[1\bar{1}2]$ direction. The thick and thin lines represent the CSL and DSC lattices of $\Sigma 11$, respectively. Reprinted with permission from Shimokawa (2010). Copyright 2010 by American Physical Society

$$b^{\text{gb}+} - b^{\text{lt}} = 2b_2^{\text{dsc}} - b^{\text{lt}}$$

$$= \frac{2}{11}[1\bar{3}\bar{1}]_{\text{II}} - \frac{1}{2}[1\bar{1}0]_{\text{II}}$$

$$= \frac{1}{22}[\bar{7}1\bar{4}]_{\text{II}} = -b_1^{\text{dsc}}. \tag{3.1}$$

This means that the residual Burgers vector at the grain boundary is the same as in the transverse DSC lattice and that there are no grain boundary dislocations that form grain boundary misorientation angles. In other words, the $\Sigma 15$ grain boundary can transition to the $\Sigma 11$ grain boundary by emitting lattice dislocations under tensile deformation. Such a dislocation emission accompanied by a grain boundary transition has also been observed in the case of a symmetrical tilt grain boundary of $\langle 110 \rangle$ (Spearot 2008). Notice that, in the case of compressive deformation at the $\Sigma 15$ grain boundary, the Burgers vectors of the lattice dislocations were in the opposite direction to b^{lt}, so a dislocation cannot be emitted from the grain boundary dislocation, as the magnitude of the residual Burgers vector becomes larger than that of the grain boundary dislocation. The grain boundary structure transition by compressive deformation occurred at the grain boundaries with a smaller misorientation angle than that of the $\Sigma 11$ grain boundary, which is the case with the $\Sigma 21$ grain boundary, where the Burgers vector of the grain boundary dislocation was $-b^{gb+}$ when the $\Sigma 11$ was the reference structure. Consequently, if the Burgers vectors of the grain boundary dislocations and lattice dislocations, which can be activated under an applied loading, are close to each other, lattice dislocations can easily be generated from the grain boundary accompanied by a grain boundary structure transition into the stable boundary, which forms a disclination dipole at the grain boundary.

3.3.3 Intragranular Crack

To investigate the disclination shielding effect on the fracture toughness, the interactions among the crack tip, dislocations, and grain boundary were simulated using molecular dynamics (Shimokawa et al. 2011). Here, we used the $\Sigma 15$ grain boundary ($|BBC|$ period), which has a high dislocation source ability in the tensile deformation described in the previous section.

3.3.3.1 Transition of the Dislocation Sources from the Crack Tip to the Grain Boundary

Figure 3.15a–c shows the evolution of the lattice defects in a model with intergranular cracks under tensile loading with a strain rate of 2×10^7 1/s at 100 K. The color of the atoms represents the rotation angle extent from the initial state. The embedded atom method proposed by Mishin et al. was adopted to simulate the atomic interactions. Mishin et al. (1999). First, since the crack tip became a strong source of stress concentration, dislocations were generated from the crack tip, as shown in Fig. 3.15a, and these dislocations penetrated into the grain boundary. These dislocations also shielded the crack tip and reduced the stress concentration capability: *dislocation shielding and grain boundary shielding*. However, as shown in Fig. 3.15b, c, a large number of dislocations were generated from the $\Sigma 15$ grain boundary due to the stress concentration caused by the lattice dislocations near the grain boundary. As shown in Fig. 3.15d, the lattice dislocations were emitted from the C structural units, which contained the grain boundary dislocations, and the grain boundary structure transitioned to the more stable $\Sigma 11$ grain boundary (B only). Also, a disclination dipole was formed at its boundary. This can be confirmed by the rotation of the region in which the dislocations were emitted, as shown in Fig. 3.15c.

3.3.3.2 Improving the Fracture Toughness by Disclination Shielding

Figure 3.16 shows the effect of the mechanical field of the disclination dipole formed at the grain boundary on the crack tip. Figure 3.16a shows the stress field of the grain boundary shielding by a lattice dislocation emitted from the crack tip, and

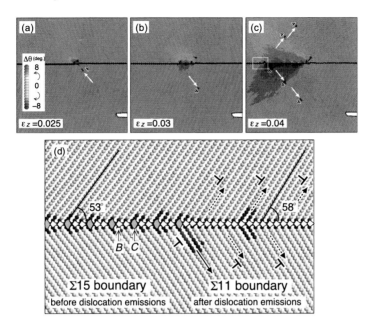

Fig. 3.15 a, b, c Translation of the dislocation source from the crack tip to the grain boundary in model A. The atomic color represents the rotation angle from the initial state of each atom with an FCC structure, $\Delta\theta$. The black atoms represent the defect structures. **d** Change in the grain boundary structures from $\Sigma 15$ to $\Sigma 11$ after the dislocation emissions from the C structural units. This atomic structure corresponds to a higher resolution image of the white box shown in (**c**). The dark and light gray atoms indicate that the nearest-neighboring atomic configurations correspond to stacking fault structures and other defects, respectively. The atoms shown in other colors form an FCC structure. Reprinted with permission from Shimokawa et al. (2011). Copyright 2011 by American Physical Society

Fig. 3.16b, c shows the stress field of the disclination dipole with the same Frank vector but with a different length a_{discli} by introducing the $\Sigma 11$ grain boundary region into the $\Sigma 15$ grain boundary. These are all unloaded conditions. With the increase in a_{discli}, a negative stress field was generated at the crack tip, indicating that the disclination dipole shielded the crack tip. To more quantitatively evaluate this effect, Fig. 3.16d shows the value of the local stress intensity factor due to the disclination dipole k_{discli}, which was normalized by the local stress intensity factor k_{gb,l_t} due to the grain boundary shielding shown in Fig. 3.16a. It was confirmed that the shielding effect of the crack tip increased with the increase in a_{discli}. Thus, the fracture toughness of the crack tip could be increased by the disclination dipole. The positional relationship between the crack tip and dislocations emitted from the grain boundary changed as a result of the dislocation motion, but the positional relationship between the disclination dipole and crack tip remained unchanged. Therefore, the dislocation emission from the grain boundary allowed the disclination dipole to grow while maintaining the plastic deformation capacity around the crack tip. It could also

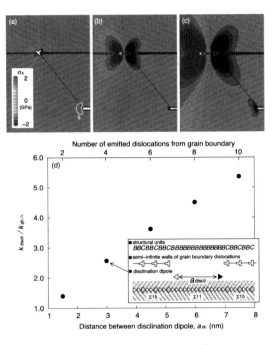

Fig. 3.16 Disclination shielding and its dependence on the distance between the disclination dipoles, a_{discli}. Distributions of σ_z by the **a** grain boundary shielding, k_{gb,l_t}, **b** disclination shielding, k_{discli} with $a_{\text{discli}} = 3$ nm, and **c** k_{discli} with $a_{\text{discli}} = 6$ nm when the applied stress σ_a was zero. **d** The normalized disclination shielding effect $k_{\text{discli}}/k_{\text{gb},l_t}$ as a function of a_{discli}. The inset figure represents the grain boundary structure through the combination of the structural units B and C, two semi-infinite walls of the grain boundary dislocations, and a disclination dipole. Reprinted with permission from Shimokawa et al. (2011). Copyright 2011 by American Physical Society

better shield the mechanical field at the crack tip and improve the fracture toughness with the progress in the plastic deformation. As a result, it can be concluded that one of the mechanisms for improving the fracture toughness of UFG materials is that grain boundaries act as a source of dislocations and create a shielding field by the disclination dipole in the grain boundary region against the mechanical field at the crack tip, which can be realized by the grain boundary transformation through dislocation emissions.

3.3.4 Intergranular Crack

The same kind of disclination shielding was also observed in the intergranular cracks, as shown in Fig. 3.13b, c (Shimokawa and Tsuboi 2015). Based on the energy equilibrium, the following equations should be satisfied for the propagation of intergranular cracks (Sutton and Balluffi 1995):

$$G_c = G > (2\gamma_s - \gamma_{gb}) + \gamma_p. \tag{3.2}$$

Here, G, γ_s, γ_{gb}, and γ_p represent the strain energy release rate, surface energy, grain boundary energy, and work done by the plastic strain near the crack tip, respectively. G is the crack extension force, and $(2\gamma_s - \gamma_{gb}) + \gamma_p$ is the crack extension resistance. Generally, since γ_p is much larger than $(2\gamma_s - \gamma_{gb})$, it can be predicted that the grain boundary fracture is not affected by the grain boundary structure. However, it has been reported that the fracture toughness is significantly affected by the segregation amount at grain boundaries (Guttmann and McLean 1979), which is contrary to the prediction of Eq. (3.2). It has been proposed that this contradiction can be resolved by considering γ_p to be a function of $(2\gamma_s - \gamma_{gb})$ (Jokl et al. 1980). In other words, the plasticity capacity near the intergranular cracks may be affected by the grain boundary characteristics. In this section, a tensile deformation analysis was conducted for the intergranular cracks of $\Sigma 15$ and $\Sigma 73$, which tend to emit dislocations under tensile loading as in the previous section, using molecular dynamic simulations to investigate the relationship between the grain boundary characteristics and fracture toughness while focusing on disclination shielding (Shimokawa and Tsuboi 2015). The analysis material was aluminum, the strain rate was 1×10^8 1/s, and the temperature was maintained at 100 K.

3.3.4.1 Propagation of the Stress Concentration Field

The results of the uniaxial tensile simulation of the analytical model with a crack on the right side of the $\Sigma 15$ grain boundary are shown in Fig. 3.17. The colors of the atoms represent the normal stress components in the loading direction. The grain boundary dislocations in front of the intergranular crack were continuously emitted from the C structural unit. When dislocations were emitted from the grain boundary, the $\Sigma 15$ grain boundary structure became $\Sigma 11$, and the grain boundary misorientation angle changed. This implies that a disclination dipole was formed at the grain boundary, as shown in Fig. 3.17c, d. As a result, the stress concentration field at the crack tip shifted to the left along the grain boundary, as if an intergranular crack was growing. In other words, the mechanical field due to the crack tip was shielded by the disclination dipole, which increased the fracture toughness. In addition, the energy of the grain boundary decreased after the dislocation emissions, meaning that the grain boundary was transitioned to one that is less likely to be the cleavage fracture.

3.3.4.2 Nanograin Formation

A crack was introduced at the left side of the $\Sigma 73$ grain boundary ($|BBBBBBBC|$ period), and the results of the uniaxial tensile analysis perpendicular to the crack plane are shown in Fig. 3.18. As dislocations were emitted from the crack tip, grain boundary dislocations in front of the crack tip were also emitted. The grain boundary

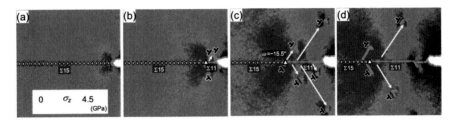

Fig. 3.17 Shift of the stress concentration ahead of the intergranular crack tip of the Σ15 boundary by the disclination shielding shown in Fig. 3.13b, which was caused by the continuous dislocation emissions from the grain boundary. Reprinted with permission from Shimokawa and Tsuboi (2015) from Elsevier

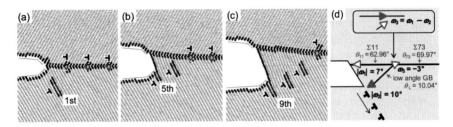

Fig. 3.18 Nanograin formation near the crack tip caused by the dislocation emission from the crack tip and the Σ73 grain boundary. **d** Schematic of the disclination dipoles ahead of the crack tip. The nanograin with a wedge disclination with ω_3 at the triple junction among the original Σ73 boundary, Σ11, and low angle grain boundary. Reprinted with permission from Shimokawa and Tsuboi (2015) from Elsevier

structure at the crack tip was branched into the Σ11 grain boundary and a small-angle grain boundary, and a new nanocrystal grain was formed in front of the crack. If the sum of the misorientation angle θ_{73} of the original grain boundary, θ_{11} of the Σ11 boundary, and θ_L of the small-angle grain boundary, 11, is not equal, there is a disclination with the Frack vector ω_3 at the triple junction, as shown in Fig. 3.18d. In the grain boundaries with $\theta > \theta_{11}$, the greater the θ of the original grain boundary, the greater the θ_L of the small-angle grain boundary formed near the crack, resulting in a larger negative Frank vector ω_3 in front of the crack that can shield the intergranular crack. Therefore, by controlling the character and strength of disclinations near the crack tip caused by the lattice defect evolution, which is closely related to the role of GBs as dislocation sources, materials with excellent strength and ductility can be obtained.

3.4 Conclusion

In this chapter, two different studies involving the disclination-mediated plastic phenomena observed in atomic simulations were reviewed to show the potential of disclination, which exposes the attractive properties of nanostructured metals. The first reviewed study discussed a grain subdivision mechanism related to the mobility of partial disclination under severe plastic deformation processes. The second one thoroughly discussed an improving fracture toughness mechanism in which the disclination shielding effect appears at the grain boundary after dislocation emission. These atomic simulations with the geometrical restrictions of boundary conditions showed the possibility of selecting a plastic deformation mode by designing structures, elements, and environments to obtain materials with excellent mechanical properties.

Acknowledgements The author acknowledges financial support from MEXT KAKENHI Grant numbers 22102002 and 22102007 (Grants-in-Aid for Scientific Research on Innovative Areas, "Bulk Nanostructured Metals").

References

Ashby MF (1970) Phil Mag 21(170):399–424
Bollmann W (1970) Crystal defects and crystalline interfaces. Springer, New York
Cheng Z, Zhou H, Lu Q, Gao H, Lu L (2018) Science 362(6414)
Chokshi AH, Rosen A, Karch J, Gleiter H (1989) Scr Metall 23(10):1679–1683
Fougere GE, Weertman JR, Siegel RW, Kim S (1992) Scripta Metallurgica et Materiala 26(12):1879–1883
Giannattasio A, Tanaka M, Joseph TD, Roberts SG (2007) Phys Scr 2007:87–90
Gleiter H (2000) Acta Mater 48(1):1–29
Gutkin MY, Ovid'ko IA (2004) Plastic deformation in nanocrystalline materials, vol 7. Springer
Gutkin MY, Ovid'ko IA, Skiba NV (2003) Acta Mater 51(14):4059–4071
Guttmann M, McLean D (1979) In: Johnson WC, Blakely JM (eds) Segregation interfacial. Am. Soc. Metals, Metals Park, OH, pp 261–347
Hansen N, Mehl RF, Medalist A (2001) Metall Mater Trans A 32(12):2917–2935
Hodge JM, Manning RD, Reichhold HM (1949) JOM 1(3):233–240
Iwahashi Y, Wang J, Horita Z, Nemoto M, Langdon TG (1996) Scripta Mater 35(2):143–146
Jokl ML, Vitek V, McMahon CJ (1980) Acta Metall 28(11):1479–1488
Lu L, Chen X, Huang X, Lu K (2009) Science 323(5914):607–610
Majumdar BS, Burns SJ (1981) Acta Metall 29(4):579–588
Meyers MA, Mishra A, Benson DJ (2006) Prog Mater Sci 51(4):427–556
Mishin Y, Farkas D, Mehl MJ, Papaconstantopoulos DA (1999) Phys Rev B Condens Matter Mater Phys 59(5):3393–3407
Mishin Y, Mehl M, Papaconstantopoulos D, Voter A, Kress J (2001) Phys Rev B 63(22):224106(1–16)
Murayama M, Howe JM, Hidaka H, Takaki S (2002) Science (New York, N.Y.) 295(5564):2433–2435
Nakashima K, Horita Z, Nemoto M, Langdon T (2000) Mater Sci Eng, A 281(1–2):82–87
Noronha SJ, Farkas D (2004) Mater Sci Eng, A 365(1–2):156–165
Raab GI (2005) Mater Sci Eng, A 410–411(May):230–233
Rice JR, Thomson R (1974) Phil Mag 29(1):73–97
Romanov AE, Vladimirov VI (1992) Dislocations in solids, vol 9, pp 191–402

Saito Y, Utsunomiya H, Tsuji N, Sakai T (1999) Acta Mater 47(2):579–583

Sawangrat C, Kato S, Orlov D, Ameyama K (2014) J Mater Sci 49(19):6579–6585

Schiøtz J, Jacobsen KW (2003) Science 301(5638):1357–1359

Shimokawa T (2010) Phys Rev B 82(17):174122(1–13)

Shimokawa T (2012)Roles of grain boundaries in the strength of metals by using atomic simulations. In: Fan H, Iwamoto N, Yuen M (eds) Molecular modeling and multiscaling issues for electronic material applications. Springer, Boston, Chap 4, pp 55–75

Shimokawa T, Kitada S (2014) Mater Trans 55(1):58–63

Shimokawa T, Tsuboi M (2015) Acta Mater 87:233–247

Shimokawa T, Kinari T, Shintaku S, Nakatani A (2006) Modell Simul Mater Sci Eng 14(5):S63–S72

Shimokawa T, Yamashita T, Niiyama T, Tsuji N (2016) Mater Trans 57(9):1392–1398

Shimokawa T, Nakatani A, Kitagawa H (2005) Phys Rev B 71(22):224110(1–8)

Shimokawa T, Tanaka M, Kinoshita K, Higashida K (2011) Phys Rev B 83(21):214113(1–13)

Spearot DE (2008) Mech Res Commun 35(1–2):81–88

Sun L, He X, Lu J (2018) NPJ Comput Mater 4(1):1–18

Sutton AP, Balluffi RW (1995) Interfaces in crystalline materials. Oxford Science, New York

Tanaka M, Fujimoto N, Higashida K (2008) Mater Trans 49(1):58–63

Tanaka M, Higashida K, Shimokawa T, Morikawa T (2009) Mater Trans 50(1):56–63

Tsuji N, Okuno S, Koizumi Y, Minamino Y (2004) Mater Trans 45(7):2272–2281

Tsuji N, Ogata S, Inui H, Tanaka I, Kishida K, Gao S, Mao W, Bai Y, Zheng R, Du JP (2020) Scripta Mater 181:35–42

Valiev RZ, Langdon TG (2006) Prog Mater Sci 51(7):881–981

Valiev RZ, Ivanisenko YV, Rauch EF, Baudelet B (1996) Acta Mater 44(12):4705–4712

Valiev RZ, Islamgaliev RK, Alexandrov IV (2000) Prog Mater Sci 45(2):103–189

Wang J, Misra A (2011) Curr Opin Solid State Mater Sci 15(1):20–28

Yinmin W, Mingwei C, Fenghua Z, En M (2002) Nature 419(6910):912–915

Zeng XH, Hartmaier A (2010) Acta Mater 58(1):301–310

Zhang Z, Vajpai SK, Orlov D, Ameyama K (2014) Mater Sci Eng, A 598:106–113

Zhou K, Nazarov A, Wu M (2006) Phys Rev B 73(4):1–11

Zhou K, Nazarov A, Wu M (2007) Phys Rev Lett 98(3):1–4

Chapter 4
Collective Motion of Atoms in Metals by First Principles Calculations

Isao Tanaka and Atsushi Togo

4.1 Introduction

The *plaston* concept has been proposed recently in order to explain interesting plastic deformation behaviors that appear in some ultrafine grained metals or bulk nanostructured metals (BNM). They show both high strength and large tensile ductility contrary to the general trade-off relationship between strength and ductility in metals (Tsuji and Ogata et al. 2020). Normal dislocation mode is predominant in the early stage of plastic deformation of ordinary metals. Sequential nucleation of different deformation modes, such as unusual dislocation modes, deformation twinning, and martensitic transformation, would induce strain-hardening ability of such BNMs, leading to high strength and large ductility.

Collective motion of atoms occurs under a stress field in materials leading to nucleate various plastic deformation modes that are not only dislocation glide but also deformation twinning and martensitic transformation. The atomic process of plastic deformation is called *plaston*. A logical way to scrutinize the collective motion

I. Tanaka (✉) · A. Togo
Center for Elements Strategy Initiative for Structural Materials (ESISM), Kyoto University, Sakyo-ku, Kyoto 606-8501, Japan
e-mail: tanaka@cms.mtl.kyoto-u.ac.jp

A. Togo
e-mail: togo.atsushi@gmail.com

I. Tanaka
Department of Materials Science and Engineering, Kyoto University, Sakyo-ku, Kyoto 606-8501, Japan

Nanostructures Research Laboratory, Japan Fine Ceramics Center, Atsuta, Nagoya 456-8587, Japan

A. Togo
Research and Services Division of Materials Data and Integrated System, National Institute for Materials Science, Tsukuba, Ibaraki 305-0047, Japan

I. Tanaka et al. (eds.), *The Plaston Concept*,
https://doi.org/10.1007/978-981-16-7715-1_4

of atoms is given by systematic calculations tracing imaginary phonon modes in deformed crystals. Although nucleation of plastic deformation modes in polycrystalline materials predominantly takes place at grain boundaries and other crystalline imperfections where atomic arrangements are irregular and stress field is complicated, we have started our calculations on perfect crystals in order to study the collective motion of atoms in a simple way. Firstly, we will show the collective motion of atoms associated with the phase transition in metallic elements (Togo and Tanaka 2015a). A simple algorithm for automated searching of the phase-transition pathway following the imaginary phonon modes is presented. Secondly, the collective motion of atoms in HCP-Ti under homogeneous shear deformation corresponding to the $\{10\bar{1}2\}$ twinning mode (Togo et al. 2020) will be shown.

4.2 Phase-Transition Pathway in Metallic Elements

The phase-transition pathway (PTP) connects structures before and after the phase transition by continuous atomic displacements and lattice deformation. In the past, PTP was searched via group-subgroup relationships of crystal structures choosing a limited number of pathways. The energy change along the PTP was not examined. Consequently, the resulted pathways were not necessarily realistic.

The PTPs can be unambiguously examined by a combination of the first principles calculations and the symmetry analysis. Although the PTP can be started from an arbitrarily chosen crystal structure, the starting crystal is better to have high symmetry and high energy. In the study described in Togo and Tanaka (2015a), we took a simple cubic (SC) structure as the start. For the first principles calculations, the plane-wave basis projector augmented wave method (Blöchl 1994) within the framework of density functional theory (DFT) as implemented in the VASP code (Kresse Non-Cryst 1995; Kresse and Furthmüller 1996; Kresse and Joubert 1999) was employed. Dynamical matrices of a given crystal were firstly obtained by the first principles phonon calculation using the phonopy code (Togo and Tanaka 2015b). The dynamical stability of the crystal was investigated by the eigenvalues of the dynamical matrices. The square root of the eigenvalue is the phonon frequency. The negative eigenvalue, i.e., the imaginary phonon frequency indicates the dynamical instability. If the sign of the instability was found, the structure was deformed along the direction indicated by the eigenvector of the unstable eigensolution that provides information of collective displacement of atoms to break the crystal symmetry minimally. In this manner, the symmetry constraint was removed, and the geometry of the structure could be further relaxed. Since it is difficult to relax the structure strictly to the local minimum of the potential energy surface at which high crystal symmetry may recover, the relaxed structure was cast in one of space group types accepting a predetermined tiny tolerance. The procedure was repeated by returning to the phonon calculations of the relaxed structure. When there were multiple instabilities, the procedure was split into multiple branches. Finally, the set of the connections produced a line diagram as shown in Fig. 4.1 for Cu and Mg.

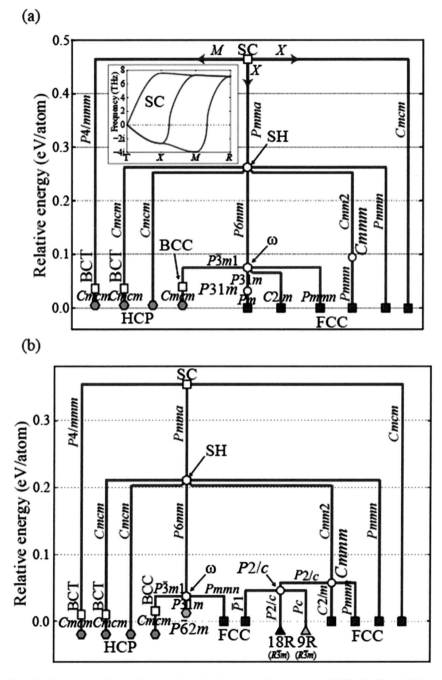

Fig. 4.1 Structure evolution diagrams showing phase-transition pathway (PTP) of **a** Cu and **b** Mg. Phonon band structure of simple cubic (SC) Cu is shown in the inset. Open and filled symbols, respectively, represent dynamically unstable and stable crystal structures. Adapted from Togo and Tanaka (2015a) with small modification (CC BY 3.0)

Instabilities of phonon were found at the wave vectors of M and X points for SC-Cu as shown in the inset (phonon band structure) of Fig. 4.1. The deformation at the M point broke the symmetry of the SC structure ($Pm\bar{3}m$) to $P4/nmm$. The twofold degenerated instability at the X point led to $Pmma$ and $Cmcm$. A calculation unit was defined as a set of procedures from one phonon calculation including relaxation of the input crystal structure to the creation of deformed crystal structures following the dynamical instabilities. The deformed crystal structures were used as the input of the next units. The next units were processed in the same way. The deformed crystal structure of $P4/nmm$ was relaxed to the body-centered tetragonal (BCT) structure. $Pmma$ and $Cmcm$ were relaxed to simple hexagonal (SH) and face-centered cubic (FCC) structures, respectively. If no instability was found in a unit, this unit ended. The unit of $Cmcm$ formed no next unit since FCC was dynamically stable for Cu. The whole procedure finished when all crystal structures of the end-point units became dynamically stable.

In Fig. 4.1, PTPs are shown by thick (blue) lines. The line was drawn only when the final structures were made by continuous atomic displacements and lattice deformation of the initial structure. The common subgroup of initial and final structures was written vertically next to the line. The energy decreased monotonously with the phase transition along the line, meaning that the transition can occur without an energy barrier. The line ended when the final structure became dynamically stable.

In this study, the supercell size was limited in order to search for simple PTPs. The explosion of the computational demands can then be avoided. Both of the 3 × 3 × 2 and 2 × 2 × 2 supercells were used to calculate force constants for the hexagonal primitive cells, and the 2 × 2 × 2 supercells were used for the other primitive cells. If the length of a supercell lattice vector exceeded 20 Å, the primitive cell was no more expanded in this direction. The elastic instability was calculated by introducing strains to the crystal structures. However, we did not find any different crystal structures from those found without considering the elastic instability. The freedom of elastic instabilities was not included for the simplicity of the algorithm.

We can find some similarities and differences between diagrams of Cu and Mg. HCP is lower in energy than FCC in Mg, which is consistent with the fact that Mg forms HCP under ordinary conditions. 9R structure ($R\bar{3}m$), known as the samarium structure, is one of the long-period-stacking (LPS) structures with ABABCBCACA stacking of the close packed planes. The diagram shows that the 9R structure is formed from the $Cmmm$ structure via $P\bar{2}/c$. The structural relationships of the FCC and 9R structures to the $Cmmm$ are shown in Fig. 4.2. Another LPS structure, 18R, is found in the diagram of Mg. Both 9R and 18R structures are experimentally known to be formed in Mg alloys.

Diagrams for two HCP metals, Ti and Hf, are displayed in Fig. 4.3. As expected, HCP shows lower energy than FCC. Two diagrams look similar. But, the path connecting SH and ω is absent in Ti, which implies that there is an energy barrier in the SH → ω path. The SH → ω path can be seen both in Cu and Mg (Fig. 4.1) and also in Hf. We made a separate set of calculations for the ω structure of 27 transition elements. (Ikeda and Tanaka 2016) ω-Ti is dynamically stable and the energy is lower than HCP. ω-Ti was experimentally reported to be existent under the external

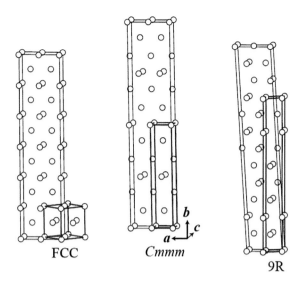

Fig. 4.2 Crystal structures of FCC, *Cmmm*, and 9R structures. Thick lines show the edges of the conventional unit cells. The conventional unit cell of the *Cmmm* structure is doubled along both of *a* and *b* axes to show the correspondence between *Cmmm* and 9R structures. Adapted from Togo and Tanaka (2015a) with small modification (CC BY 3.0)

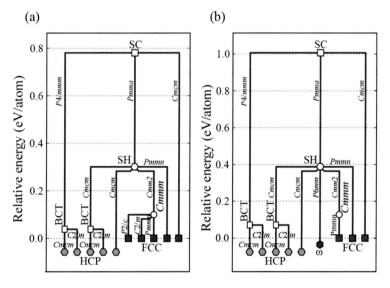

Fig. 4.3 Structure evolution diagrams of **a** Ti and **b** Hf. Adapted from Togo and Tanaka (2015a) with small modification (CC BY 3.0)

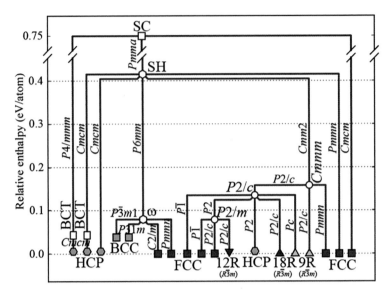

Fig. 4.4 Structure evolution diagram of Cu at 30 GPa. Adapted from Togo and Tanaka (2015a) with small modification (CC BY 3.0)

pressure of 2 GPa at 293 K (Sikka et al. 1982). The critical pressure decreases with decreasing temperature by approximately $dT/dp = 100$ K/GPa (Sikka et al. 1982), which suggests that ω is more stable than HCP at temperatures <90 K. However, a large hysteresis was experimentally reported for the transition between HCP and ω under pressure. The HCP → ω transition has not yet been reported experimentally under the ordinary pressure. The presence of a large energy barrier between SH and ω is implied by these experimental results. The absence of the SH → ω path in Fig. 4.3a is consistent with the experimental reports.

Pressure effects on the structure evolution diagram can also be examined in the same way as described above. Figure 4.4 shows the result for Cu under 30 GPa of hydrostatic pressure. Note that the vertical axis shows relative enthalpy here. As compared to the zero-pressure case shown in Fig. 4.1a, the number of dynamically stable phases increased under 30 GPa. BCC-Cu became dynamically stable. The formation of BCC-Cu was experimentally reported in the initial process precipitation in Fe-Cu alloys (Othen et al. 1994). Three LPS structures, i.e., 9R, 12R, and 18R appeared under 30GPa, similar to the case of Mg (0 GPa). The 9R-Cu was experimentally reported as a thin layer of the Σ3 grain boundary in Cu (Wolf and Ernst 1992) and in severely deformed Cu single crystals (Dymek and Wróbel 2003).

4.3   HCP-Ti Under Shear Deformation Along Twinning Mode

Figure 4.5a shows the conventional unit cell of HCP-Ti. In order to represent the $\{10\bar{1}2\}$ twinning mode, the unit cell is retaken by using η_1 and η_2 directions and K_1 and K_2 planes as shown in Fig. 4.5b. P (green) is the shear plane. Symmetry constraints are given to distinguish four structures, i.e., the parent, sheared-parent, shuffling, sheared-twin, and twin structures. Three structures are shown in Fig. 4.5c–e. The parent and twin structures have the space group type of $P6_3/mmc$. The sheared-parent structure has the space group of $C2/m$ in general. It becomes $Cmcm$ only when the sheared lattice becomes the same lattice as the parent in a different orientation. The shuffling structure appears during the transformation from the sheared-parent to sheared-twin structures.

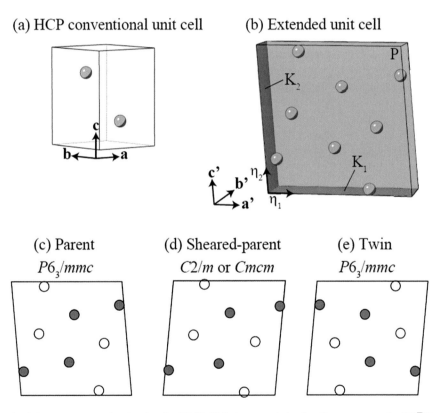

Fig. 4.5   a The conventional unit cell of HCP-Ti. **b** An extended unit cell to represent the $\{10\bar{1}2\}$ twinning mode of HCP-Ti. The twinning mode is characterized by K_1 (red) and K_2 (blue) planes and η_1 and η_2 directions. Structures of an extended unit cell of **c** parent, **d** sheared-parent, and **e** twin. The atoms with the same symbols (open or filled) are located on the same plane parallel to the plane P. Two planes with the open and filled symbols are separated by $0.5b'$ each other. Adapted from Togo et al. (2020) with small modification (CC BY 3.0)

Atomic positions of the sheared-parent structures were optimized under the symmetry constraint. Shears were sampled to 21 points within $0 \leq s/s_t \leq 1$, where s/s_t denotes homogeneous shear, s, normalized by the twinning shear, s_t. In order to introduce homogeneous shear, the positions of atoms have to be properly set since there are degrees of freedom to relax atomic positions under symmetry constraints. We defined atomic displacement $u(s/s_t)$ as the difference between the atomic positions after and before the relaxation. The initial position before the relaxation was given keeping the crystallographic coordinates of the HCP unit cell. The displacement distances $|u(s/s_t)|$ with respect to s/s_t are shown in Fig. 4.6. Since $|u(s/s_t)|$ of all atoms in the unit cell at given s/s_t are the same, only one value at each s/s_t is shown. In the inset, the directions of the displacements are shown. In this figure, only two directions that are directed in the opposite directions can be seen. The atoms having the same displacement directions are symmetrically equivalent by the lattice translation. The displacement distance $|u(s/s_t)|$ increases with s/s_t. Even at $s/s_t = 1$, $|u(s/s_t)|$ is only 4% of the nearest neighbor distance. Therefore, for the schematic analysis, this optimization process is unimportant, although it is necessary to perform phonon calculations accurately.

In Fig. 4.7, the electronic total energy of the sheared-parent structure is shown as a function of the homogeneous shear s/s_t. With increasing the shear, the energy increases harmonically. According to the definition of the deformation twinning, the sheared-twin structures should have the same energy curve with respect to $1 - s/s_t$, which is drawn as the mirror image of the energy curve of the parent (dotted curve in Fig. 4.7). The presence of the crossing of the energy curves at $s/s_t = 0.5$ means that their energy surfaces are disjoined under the symmetry constraint of $C2/m$.

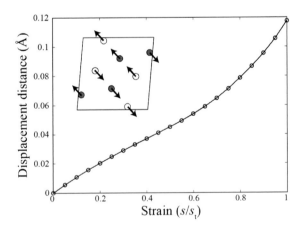

Fig. 4.6 Displacement distance, $|u(s/s_t)|$, for the $\{10\bar{1}2\}$ twinning mode as a function of s/s_t. In the inset, the sheared-parent structure at $s/s_t = 1$ before the optimization is shown. The arrows show the approximate directions of the displacements. Adapted from Togo et al. (2020) (CC BY 3.0)

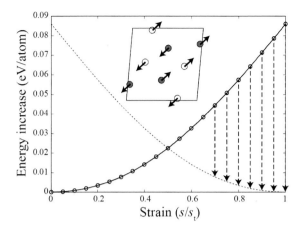

Fig. 4.7 Energy increases per atom of the sheared-parent structures for the $\{10\bar{1}2\}$ twinning mode. The dotted curve is the mirror image of the energy curve of the sheared-parent structure representing the energy increases of the sheared-twin structures. The vertical dashed arrows indicate energies released by the transformation from the sheared-parent to the sheared-twin structures. The inset shows the sheared-parent structure at $s/s_t = 1$. The displacement directions indicated by the eigenvector of the imaginary phonon mode at $M'(1/2, 0, 0)$ are shown by thick arrows. Adapted from Togo et al. (2020) (CC BY 3.0)

The sheared-parent structure transforms to the sheared-twin structure spontaneously by breaking the symmetry constraint at $s/s_t > 0.5$. The phonon band structures at six s/s_t are shown in Fig. 4.8. One phonon mode at $M'(1/2, 0, 0)$ clearly shows the imaginary frequency at large shears, which implies that spontaneous structural transformation should occur by breaking the symmetry. By plotting the squared frequencies at $M'(1/2, 0, 0)$ with respect to the homogeneous shear s/s_t, the critical shear to induce the spontaneous structural transformation can be estimated to be $s/s_t \sim 0.68$.

The eigenvector of the soft phonon mode at $M'(1/2, 0, 0)$ as illustrated in the inset of Fig. 4.7 indicates the information on the collective atomic displacements to properly break the symmetry of the crystal. The arrows indicate the direction of displacement of the atoms with the same relative amplitude. The breaking of symmetry expands the primitive cell twofold. The space group type then becomes $P2_1/c$ (No. 15). The sheared-parent structure models were generated by introducing minimum finite displacements along the directions as described above, so that the first principles calculation code, VASP, can properly detect the broken symmetry. Then, the first principles calculations were made to optimize these structure models at given homogeneous shears fixing their basis vectors. After the structure optimizations, the sheared-twin structures at $s/s_t = 0.7$ to 0.95 and the twin structure at $s/s_t = 1$ were obtained. This is the shuffling that we consider. The atomic displacements by this transformation at $s/s_t = 1$ are shown by short black arrows in Fig. 4.9. The displacement distance is roughly constant by ~0.5 Å (17% of the interatomic distance) at all s/s_t larger than the critical shear. Four atoms are involved in the

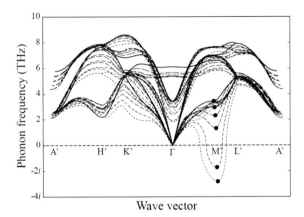

Fig. 4.8 A series of phonon band structures corresponding to the $\{10\bar{1}2\}$ twinning mode at s/s_t = 0, 0.2, 0.4, 0.6, 0.8, and 1. Those at different s/s_t are drawn by the different line styles from solid to shorter dashed lines. In vertical axis, phonon frequency lower than zero denotes imaginary frequency. Wave vectors are labeled as A′(0, 0, 1/2), H′(1/3, 1/3, 1/2), K′(1/3, 1/3, 0), Γ(0, 0, 0), M′(1/2, 0, 0), and L′(1/2, 0, 1/2). Although the homogeneous shear breaks the crystal symmetry, we adopt the same labels for points in the Brillouin zones for easy comparison. Since the shapes of the Brillouin zones are different for the different shears, the positions of the reciprocal points in the Cartesian coordinates measured from the Γ points disagree slightly. The filled circles at the M′ points indicate the soft phonon modes that have the eigenvector drawn in the inset of Fig. 4.7. Adapted from Togo et al. (2020) (CC BY 3.0)

collective displacement as the minimum unit. Two types of parallelogram units by alternately changing their rotation directions depicted by red circular arrows are arranged to fill the crystal structure. The spaces surrounded by these rotating units behave as if they were breathing. In this manner, the internal structural distortion is considered to be minimized. This figure is similar to the one obtained by Crocker and Bevis reported in 1970 (Crocker and Bevis 1970) for the shuffling of the $\{10\bar{1}2\}$ twinning mode from the crystallographic discussion. In this study, we found that the shuffling necessarily occurs at a shear larger than the critical value based on the first principles calculations.

The vertical dashed arrows in Fig. 4.7 correspond to the energy between the sheared-parent and sheared-twin structures, which is released by the shuffling. The presence of a potential energy barrier may prevent the initiation of the shuffling between $s/s_t = 0.5$ and the critical shear $s/s_t \sim 0.68$. When crystalline imperfections are present, as in a real material, the potential energy barrier may be lowered locally, and the shuffling initiated from the imperfection may propagate over the macroscopic range to result in the macroscopic twinning. Once the shuffling is locally initiated, the sheared-parent may relax toward the twin instantaneously. Although it would be difficult to observe the initiation of the twin at an atomic level, evidence of the shuffling mechanism may be experimentally obtained by measurement of the frequency change of the characteristic phonon mode at $s/s_t < 0.5$ by inelastic neutron or X-ray scattering techniques.

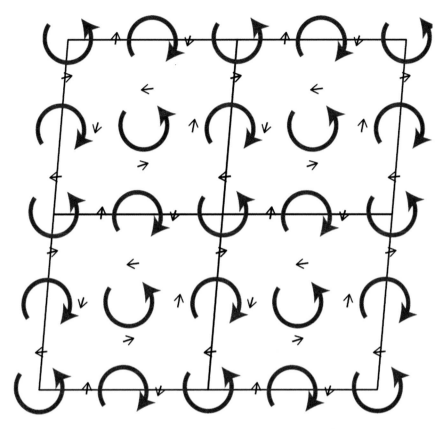

Fig. 4.9 Displacements of atoms induced by the structural relaxation from the sheared-parent structure (Fig. 4.5d) to the twin structure (Fig. 4.5e) depicted as the short black arrows. The circular red arrows depict how the displacements are arranged as units. Adapted from Togo et al. (2020) (CC BY 3.0)

Acknowledgements This work was supported by the Elements Strategy Initiative for Structural Materials (ESISM) of MEXT (Grant number JPMXP0112101000).

References

Blöchl PE (1994) Phys Rev B 50:17953
Crocker AG, Bevis M (1970) In: Jaffee RI, Promisel NE (eds)The science, technology and application of titanium. Pergamon, pp 453
Dymek S, Wróbel M (2003) Mater Chem Phys 81:552
Ikeda Y, Tanaka I (2016) Phys Rev B 93:094108
Kresse G, Furthmüller J (1996) Comput Mater Sci 6:15
Kresse G, Joubert D (1999) Phys Rev B 59:1758

Kresse G Non-Cryst J (1995) Solids 193:222
Othen PJ, Jenkins ML, Smith GDW (1994) Philos Mag A 70:1
Sikka SK, Vohra Y, Chidambaram R (1982) Prog Mater Sci 27:245
Togo A, Inoue Y, Tanaka I (2020) Phys Rev B 102:024106
Togo A, Tanaka I (2015a) Phys Rev B 87:184104
Togo A, Tanaka I (2015b) Scr Mater 108:1
Tsuji N, Ogata S et al (2020) Scr Mater 181:35
Wolf U, Ernst F et al (1992) Philos Mag A 66:991

Chapter 5
Descriptions of Dislocation via First Principles Calculations

Tomohito Tsuru

5.1 Introduction

Dislocation, a line defect in which atoms are out of the lattice position in the crystal structure, is one of the most representative "plastons". A mathematical description of the elastic field of the singularity created by cutting and shifting a continuous body was first developed by Volterra in 1907 (Volterra 1907). Since Taylor, Orowan, and Polanyi predicted the existence of dislocation to explain plastic deformation (Taylor 1934; Orowan 1934; Polanyi 1934), dislocations have been regarded as the most important lattice defects in plastic deformation, especially in metallic materials. The fundamental properties of the dislocation core have a dominant influence on the intrinsic ductility or brittleness of materials. The interaction between dislocations and other crystal defects plays a critical role in determining the mechanical properties of metals. The classical strengthening mechanism was developed by this major premise, and the mechanical properties of metals have been developed by understanding and controlling the dislocation behavior.

While continuum theory expresses an excellently long-ranged elastic field of a dislocation, it breaks down owing to the singularity near the dislocation core. With improvements in computer technology, atomic scale simulations such as molecular dynamics have been implemented to explore the relation between the dynamic

In this Chapter, DFT calculations were carried out using the Vienna ab initio simulation package (VASP) (Kresse and Hafner 1993; Kresse and Furthmuller 1996) with the Perdew–Burke–Ernzerhof generalized gradient approximation exchange–correlation density functional (Perdew et al. 1992). The Brillouin-zone k-point samplings were chosen using the Monkhorst–Pack algorithm (Monkhorst and Pack 1976). The planewave energy cutoff was set at 400 eV. The fully relaxed configurations were obtained by the conjugate gradient method when the energy norm of all the atoms converged to better than 0.005 eV/Å. Simulations were performed on the large-scale parallel computer system with SGI ICE X at JAEA.

T. Tsuru (✉)
Nuclear Science and Engineering Center, Japan Atomic Energy Agency, Tokai, Ibaraki, Japan
e-mail: tsuru.tomohito@jaea.go.jp

© The Author(s) 2022
I. Tanaka et al. (eds.), *The Plaston Concept*,
https://doi.org/10.1007/978-981-16-7715-1_5

behavior of defects and the mechanical response. Some examples of atomistic simulations for collective defect behavior were mentioned in the previous section. Recently, a dislocation core structure can be captured directly by first principles calculations. Several approaches for first principles calculations of dislocations and dislocation-related properties are described in detail in this section.

5.2 Stacking Fault Energy

The atomic image of a dislocation core is described by local disregistry, where misfit energy is introduced by the local interfacial misfit energy as well as the elastic strain energy. The misfit energy $\gamma(\mathbf{u})$ is described by periodic potential in terms of relative slip displacement, which is expressed by its sinusoidal shape potential in the classical model. Atomistic simulation, such as molecular dynamics and first principles, allows calculation of the misfit potential directly. Precise calculations of stacking fault (SF) energy can help to better understanding slip behavior in actual materials. The generalized SF energy (γ surface), which was first introduced by Vítek (Vítek 1966, 1968), is the energy difference in misfit energy in terms of two-dimensional displacement along the slip plane. Atomic models for SF energy in FCC metals are simply prepared as in Fig. 5.1, where parallel translation is applied to the upper half region. The energy of the transition during slip motion is evaluated using first principles calculations. The atomic relaxation is considered only in the normal direction to the slip plane and the other degree of freedom corresponding to the direction of the applied displacement is constrained. The difference in terms of displacement corresponds to the generalized SF energy according to following relation, where A is the area of the slip plane.

$$\gamma(\mathbf{u}) = \frac{E(\mathbf{u}) - E_0}{A}, \tag{5.1}$$

In FCC metals, the displacement is generally applied along the $[1\bar{1}0]$ and $[11\bar{2}]$ the directions defined as x and y. The Burgers vector of perfect dislocation corresponds to $a/2[1\bar{1}0]$, along which the SF energy is not minimum. Instead, the SF energy along the Burgers vector of partial dislocations is equivalent to the minimum energy path for parallel shift in the slip plane, which is important for actual slip behavior in FCC metals.

The SF energy curves for typical FCC and BCC metals calculated by first principles calculations are summarized in Fig. 5.2. The other results obtained by several empirical potential and experiments are shown as a comparison. γ_{SF} corresponds to the energy when the displacement is equivalent to the partial dislocation, shown in the right image in Fig. 5.1. The value is important in determining the stable dislocation core structure and the width between leading and trailing partial dislocations in FCC metals. Unstable SF energy γ_{us} and the maximum gradient of the generalized SF energy are also important because the local atomic array needs to overcome

Fig. 5.1 Atomic models for SF energy calculation in FCC metals

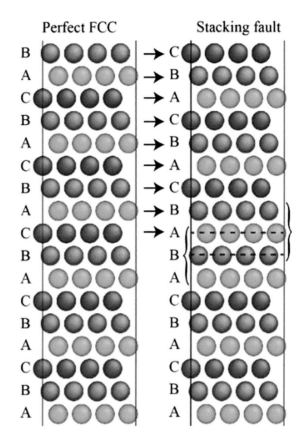

the energy barrier. First principles calculations satisfactorily reproduce the experimental values. Moreover, first principles calculations are superior in predicting the SF energy of transition states during the whole slip motion. Even though it is generally difficult for an empirical potential composed of a two-body functional to reproduce high SF energy, some EAM potentials have been found to demonstrate excellent characteristics to describe the dislocation core in FCC metals.

In HCP metals, it is more effective to understand the SF energy because the slip behavior is more complicated than that of cubic crystals. In general, plastic deformation in HCP metals is caused by dislocation of the slip and twinning deformation modes. Focusing on the slip mode, <a> and <c + a> dislocations can glide in basal, prismatic, and pyramidal planes. All possible slip systems are given in Fig. 5.3, where <a> dislocation can glide in basal, prismatic and first-order pyramidal planes, and <c + a> dislocation in first- and second-order pyramidal planes. An overview of the generalized SF energy on each plane is given in Fig. 5.4, where Mg was chosen as a typical example of HCP metals. Several tens of node points on each slip plane are chosen for the sampling data and the energy was evaluated by first principles calculations. The direction and SF energy along the minimum energy path (shown by the

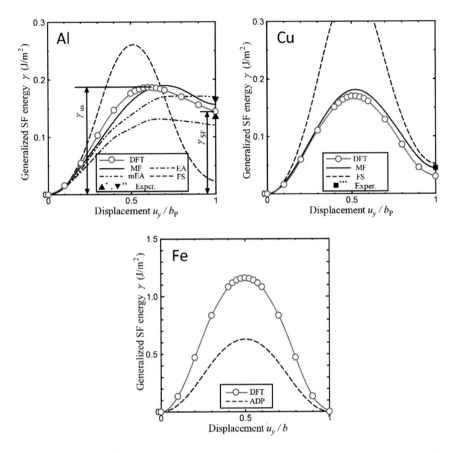

Fig. 5.2 Generalized SF energy curves along the Burgers vector in Al, Cu and Fe, calculated by first principles calculations and some empirical potentials. MF, EA, mEA, FS, and ADP correspond to the empirical potentials proposed by Mishin–Farkas (Mishin et al. 1999, 2001), Ercolessi–Adams (Ercolessi and Adams 1994), modified EA (Liu et al. 1996), Finnis–Sinclair (Finnis and Sinclair 1984), Angular dependent potential (Mishin et al. 2005). *, ** and *** are experimental values cited by the literature (Dobson et al. 1967; Murr 1975; Hirth and Lothe 1982)

red arrows) on the slip plane is the most important for determining the dislocation motion because the direction corresponds to the Burgers vector of each slip plane. In case of the basal slip, the <a> dislocation tends to dissociate into a partial dislocation and the Burgers vectors are given as those of two partial dislocations according to the minimum energy path. However, <a> dislocation in the prismatic plane does not dissociate into partials.

The basal <a> , prismatic <a> , pyramidal I <a> , pyramidal I <c + a> , and pyramidal II <c + a> slip systems can be considered as the only possible slip systems except for twin modes. Figure 5.5 shows the energy difference along the minimum energy path of all slip systems in Mg, Zn, Ti, and Zr as typical HCP metals. The c/a ratio of these HCP metals, calculated by first principles, are 1.613, 1.886, 1.582, and

Fig. 5.3 Possible slip
systems of <a> and <c + a>
dislocations in basal,
prismatic, first- and
second-order pyramidal
planes. The red arrow
indicates the direction of
each Burgers vector

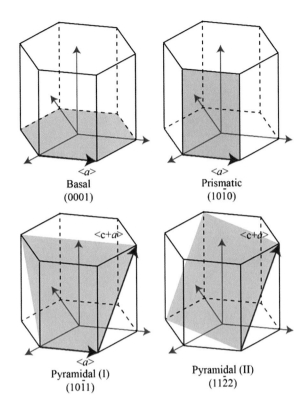

Basal
(0001)

Prismatic
(10$\bar{1}$0)

Pyramidal (I)
(10$\bar{1}$1)

Pyramidal (II)
(11$\bar{2}$2)

1.598 for Mg, Zn, Ti, and Zr, respectively. In the case of Mg and Zn, the SF energy
of the <a> slip in the basal plane is much lower than those of any other non-basal slip
planes, and the basal <a> slip is a primary slip system in these metals. In contrast, the
SF energy of the <a> slip in the prismatic plane is the lowest and the prismatic <a>
slip is the primary slip system. These tendencies can be understood using structural
information, that is, c/a is among the simplest indicators that provides a favorable slip
plane and direction. The SF energy in the basal plane tends to decrease when the c/a
ratio is larger than the ideal c/a ratio. On the contrary, the prismatic <a> becomes the
primary slip system when the ratio is smaller than the ideal. However, the SF energy
is more important for understanding the detailed information of the slip system in
HCP metals.

5.3   Analytical Description of Dislocations: Peierls–Nabarro Model

As discussed above, the SF energy calculation provides useful information for the
dislocation core and motion. However, the SF energy can only give an interfacial

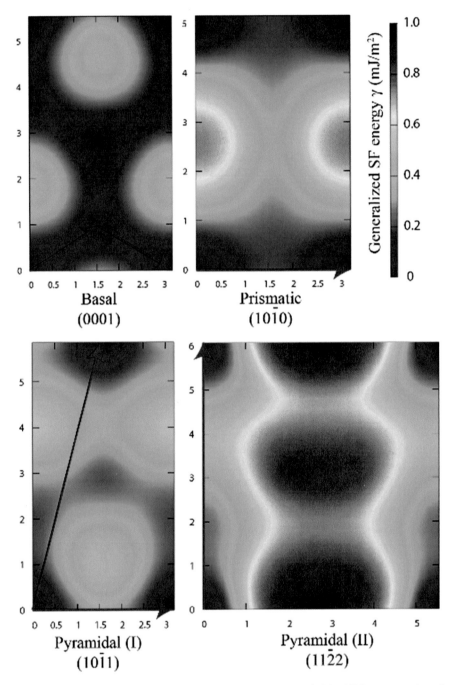

Fig. 5.4 Generalized stacking fault energies on a possible slip plane in Mg. All data were evaluated by first principles calculation. The minimum energy path is indicated by the red arrow, which corresponds to the direction of each Burgers vector

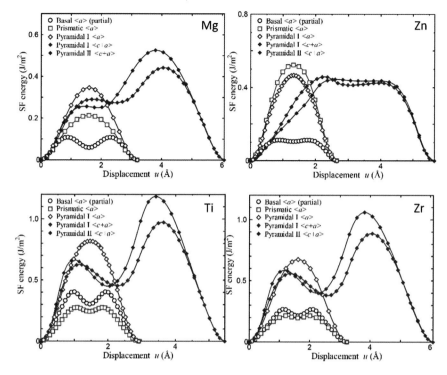

Fig. 5.5 Generalized stacking fault energies along the minimum energy path in all possible slip planes in Mg, Zn, Ti, and Zr. Here, the basal <a> , prismatic <a> , pyramidal I <a> , pyramidal I <c + a>, and pyramidal II <c + a> slip systems are considered as the possible slip systems

misfit energy of the dislocation core, which is not enough to capture a comprehensive picture of the dislocation because the elastic strain energy around the dislocation core should be considered. The self-energy of the dislocation should be described by a combination of the contributions of both the elastic energy generated by the local disconnection of the core region and the interfacial misfit energy. Peierls (Peierls 1940) and Nabarro (Nabarro 1947) first considered a simplified description of the dislocation core in a two-dimensional lattice as shown in Fig. 5.6 and proposed the Peierls–Nabarro (PN) model. Here, the dislocation core can be expressed by the two elastic semi-infinite bodies and the non-linear interplanar misfit at the interface of two bodies.

In the PN model, the relative displacement between two lattice planes at the interface is defined by the lattice mismatch (disregistry), and the disregistry is allowed to have a distribution of displacement along the direction of Burgers vector. The total displacement of the disregistry can be considered using a set of local components of displacement $\delta(x)$, which corresponds to the dislocation segment with an infinitesimal Burgers vector. The local gradient generated by the disregistry is defined by dislocation core density $\rho(x)$ as follows:

Fig. 5.6 Schematic image of
a pure edge dislocation core
in a two-dimensional lattice

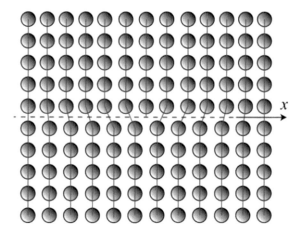

$$db(x) \equiv \rho(x)dx, \quad \rho(x) = \left(\frac{d\delta(x)}{dx}\right). \tag{5.2}$$

Here, the integral along the x direction within the infinite region is equal to the Burgers vector. The following PN equations hold when the stress generated by the disregistry is balanced against the interfacial restoring force F.

$$\frac{K}{2\pi} \int_{-\infty}^{\infty} \frac{\rho(x')}{x - x'} dx' = F(\delta(x)), \tag{5.3}$$

where K is the energy factor corresponding to the elastic coefficient and given for isotropic media (Foreman 1955).

$$K = \frac{\mu}{2\pi} \left(\frac{\sin^2 \theta}{1 - \nu} + \cos^2 \theta\right), \tag{5.4}$$

where μ and ν are the shear modulus and Poisson's ratio, respectively. θ is the angle between the dislocation line and Burgers vector. This PN equation is solved analytically on the assumption that the restoring force can be described by an analytic function such as the sinusoidal function. Similarly, the total energy of dislocation can be expressed by the local elastic interaction generated by the infinitesimal dislocation segments, long-range elastic interaction, and misfit potential at the interface of two bodies.

$$U_{\text{tot}}[\rho(x)] = U_{\text{elastic}} + U_{\text{misfit}} + U_{\text{long}}, . \tag{5.5}$$

$$U_{\text{elastic}} = -\frac{K}{4\pi} \int_{-\infty}^{\infty} \int_{-\infty}^{\infty} \rho(x)\rho(x') \ln|x - x'| dx dx'. \tag{5.6}$$

$$U_{\text{misfit}} = \int_{-\infty}^{+\infty} \gamma(\delta(x)) dx. \tag{5.7}$$

The long-range interaction is given analytically by the classical theory of dislocations. Accordingly, the total energy of the dislocation can be expressed by a functional of the disregistry and its derivative, that is, dislocation core density. Therefore, a stable configuration of a dislocation is obtained using the variational of this energy functional. Joós and Duesbery developed an analytical model based on the PN model to estimate the Peierls stress for wide and narrow dislocations by introducing the generalized SF energy, its maximum gradient, and the dislocation half-width (Joós and Duesbery 1997). It should be noted that the Peierls stress evaluated by the analytical function based on the PN model is solved on the assumption that the dislocation is dissociated along the one-dimensional direction only and that the estimation depends on the accuracy of the maximum restoring force.

Recently, a sophisticated model of the semi-discrete variational (SV) PN model (Bulatov 1997; Lu et al. 2000) was developed, which provides information on the three-dimensional local structure of a dislocation core and the Peierls stress through discretized atomic rows and the reliable GSF energy surface (so-called γ surface) determined by DFT. In the SVPN framework, the total fault energy of a dislocation under external stress can be described similarly by $U_{tot} = U_{\text{elastic}} + U_{\text{misfit}} + U_{\text{stress}}$, where long-range interaction is excluded. The first term is the configuration-dependent elastic energy is as follows:

$$U_{\text{elastic}} = \sum_{i,j} \frac{1}{4\pi} \chi_{ij} [K_e(\rho_i^{(1)}\rho_j^{(1)} + \rho_i^{(2)}\rho_j^{(2)}) + K_s\rho_i^{(3)}\rho_j^{(3)}], \tag{5.8}$$

where, $\rho_i^{(1)}$ and $\rho_i^{(2)}$ are related to the edge component and $\rho_i^{(1)}$ is related to the screw component, K_e and K_s are associated with the energy factor evaluated by Eq. (5.3), and χ_{ij} is defined in the literature (Bulatov 1997; Lu et al. 2000). The second term is the interfacial energy described by the total sum of nodes:

$$U_{\text{misrit}} = \sum_i \Delta x \gamma(\delta_i). \tag{5.9}$$

The third term represents the work done by the plastic deformation through the dislocation glide. Here, all terms can be expressed as the discretized the functions of disregistry vector and the core density of dislocation. According to the displacement of each atomic row x_i, the disregistry and core density are related to $\rho_i = (\delta_i - \delta_{i-1})/(x_i - x_{i-1})$, which is associated with the local gradient of relative displacement. In accordance with the classical PN model, the stable configuration of the dislocation core can be evaluated numerically by minimizing the variational function to find the stable configuration of the disregistry (or the dislocation core density).

The basal and prismatic <a> dislocations in HCP Mg alloys are taken as an example of good application of the SVPN framework. First, atomic configurations

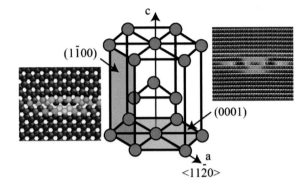

Fig. 5.7 Unit cell of hexagonal crystal and the core structures of the basal and prismatic <a> dislocations from molecular statics simulations (Tsuru et al. 2013). Reprinted from J. Phys: Condens. Matter., vol. 25, Tsuru et al., Fig. 1a, Copyright (2020), with permission from IOP

of both dislocations were investigated by atomistic simulations. The crystallographic orientation and corresponding atomic configurations are shown in Fig. 5.7. A fully-atomistic molecular statics simulation was performed using the EAM potential proposed for Mg (Sun et al. 2006). The atoms were visualized using the ATOMEYE visualization software (Li 2003). The fully-atomistic simulations confirmed that a basal dislocation tends to dissociate into two partial dislocations with wide stacking fault while the core width of a prismatic dislocation is narrow. The energy differences with respect to the in-plane displacement along the minimum energy path (slip direction) in the basal and prismatic planes evaluated by the density functional theory (DFT) calculation and EAM potential are shown in Fig. 5.8. The DFT data were the same as shown in Fig. 5.5 while the energy landscape along the slip direction of both the basal and prismatic planes had similar characteristics. It is noted that the direction of minimum energy path in the prismatic plane varied slightly (Tsuru et al. 2013), and the differences of the generalized SF energy as well as the elastic properties reflected the dislocation core structure.

Fig. 5.8 Generalized SF energy along the minimum energy path evaluated by DFT and the EAM potential (Tsuru et al. 2013). Reprinted from J. Phys: Condens. Matter., vol. 25, Tsuru et al., Fig. 1 (c), Copyright (2020), with permission from IOP

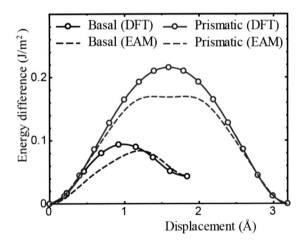

The energy factors for the pure edge and screw dislocations in the basal and prismatic planes of magnesium were evaluated directly by DFT and the EAM potential, as summarized in Table 5.1. The anisotropic shear modulus is the value of the slip direction, i.e., C_{44} and C_{66}. Poisson's ratio is assumed to be evaluated by orthogonally transformed anisotropic elastic constants through the direction cosine and the Voigt average.

All material properties of Mg are applied to SVPN analysis. The core densities of the basal and prismatic <a> dislocation, calculated by SVPN analysis combined with DFT calculations and the EAM potential, are summarized in Fig. 5.9. Two peaks

Table 5.1 Energy factors calculated by first principles and the EAM potential for the evaluation of the elastic energy of the dislocation core (Tsuru et al. 2013). Reprinted from J. Phys: Condens. Matter., vol. 25, Tsuru et al., Table 1 (c), Copyright (2020), with permission from IOP

	Basal <a>		Prismatic <a>	
	K_e (GPa)	K_s (GPa)	K_e (GPa)	K_s (GPa)
DFT	3.88	2.93	4.10	2.98
EAM	2.49	2.03	4.55	3.52

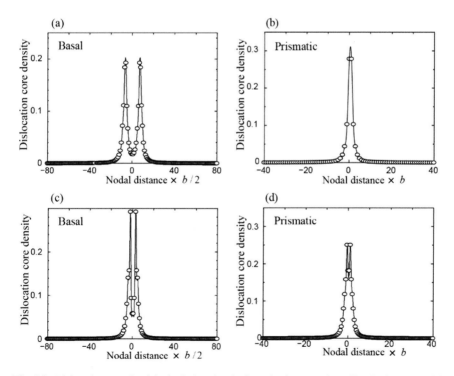

Fig. 5.9 Dislocation core densities in the basal and prismatic planes, evaluated by the SVPN model with (a, b) using DFT, and (c, d) using the EAM potential (Tsuru et al. 2013). Reprinted from J. Phys: Condens. Matter., vol. 25, Tsuru et al., Fig. 2, Copyright (2020), with permission from IOP

of the core density of the basal dislocation represent the dissociated two partial dislocations associated with the low stacking fault energy of Mg. The difference in the width of extended dislocation between DFT and EAM are affected by the meaningful difference of the energy factor, as shown in Table 5.1. The core density of dislocation in the prismatic plane evaluated by DFT distributes sharply while that of EAM slightly dissociates into two peaks, which is definitely reflected by the difference of generalized SF energy. The Peierls stresses were 1 MPa and 49 MPa for the basal and prismatic <a> dislocations, respectively. Additionally, as an example, SVPN analysis is applied to the prismatic <c> dislocation, which is not a slip system of HCP metals. SVPN analysis confirmed that the Peierls stress of the prismatic <c> dislocation was extremely high (1.49 GPa).

The non-basal slip is thought to be key to improving the ductility in Mg alloys. Thus, the effect of the alloying element is the subject of research and interest. The SVPN model has the potential to capture how alloying elements influence the dislocation core structure and the Peierls stress. As a typical example, the effect of yttrium solution on the dislocation core structure and the critical stress of the motion of prismatic <a> dislocation (i.e., Peierls stress) were discussed here. When an isolated solute atom was substituted for a magnesium atom on the slip plane, the influence of the solution element on the generalized SF energy and the restoring force are shown in Fig. 5.10. The effect of yttrium solution at different concentrations was investigated, and aluminum was chosen for comparison, which is a typical solute element of magnesium alloys such as the AZ31 alloy. The restoring force was defined as the gradient of the GSF energy along the minimum energy path, and its maximum value, denoted by τ_{max}, is related to the dislocation half-width through the analytical PN model (Joós and Duesbery 1997). The energy difference was remarkably reduced by yttrium solution, and in this regard the maximum restring force fell to one-half that of pure magnesium. However, aluminum solution raises the energy gradient and the restoring force, and zinc solution was found to show a tendency similar to that of aluminum. The dislocation core structures in case that yttrium is substituted in the slip plane, was calculated by the SVPN analysis as shown in Fig. 5.11. The dislocation tends to spread as the solute concentration increased because the contribution of the elastic interaction became relatively large owing to the reduction of the gradient of the GSF energy. The resultant Peierls stresses were reduced to 14 MPa and 8 MPa for 11 at.% and 25at.% yttrium solution concentration, respectively. Thus, yttrium solute results in the alteration of the dislocation core structure and lubrication of the dislocation motion (Tsuru et al. 2013).

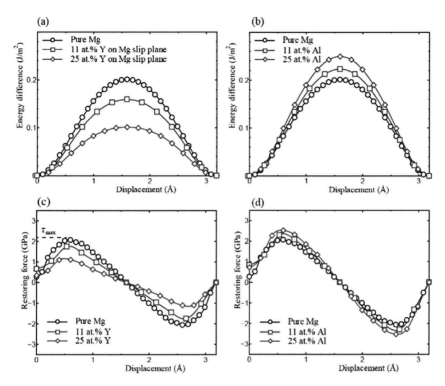

Fig. 5.10 Stacking fault energies and restoring forces in the case where a solution element is placed on the slip plane compared with those of pure magnesium. **a** and **c** correspond to yttrium solution. **b** and **d** correspond to aluminum solution (Tsuru et al. 2013). Reprinted from J. Phys: Condens. Matter., vol. 25, Tsuru et al., Fig. 3, Copyright (2020), with permission from IOP

Fig. 5.11 Effect of yttrium solution at several solution concentrations on the dislocation core density (Tsuru et al. 2013). Reprinted from J. Phys: Condens. Matter., vol. 25, Tsuru et al., Fig. 4, Copyright (2020), with permission from IOP

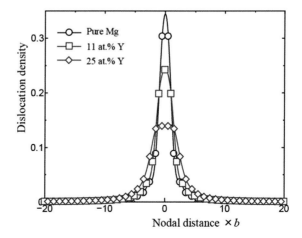

5.4 First Principles Calculations of a Dislocation Core

5.4.1 Atomic Modeling of a Dislocation Core

The effect of solute atoms on the dislocation motion is characterized by not only the SF energy but also by the elastic and chemical interactions around the dislocation core. Recently, various defect structures, such as the vacancy impurity cluster and grain boundary, can be treated directly by first principles calculations, and many fundamental properties have been extracted. However, band structure calculations, i.e., the DFT calculations based on plane-wave basis sets for the wave function of electrons, are generally solved under the period boundary condition, in which special attention is required in handling defect structures in a small supercell (up to several hundreds of atoms) because the elastic field of defects is generally long-ranged. According to the linear elasticity theory, the elastic field of a vacancy and volumetric defect, such as an impurity cluster, decays with $1/r^3$ (the factor of distance). The stress field of the coincidence site lattice (CSL) grain boundary is also short-ranged. As a result, these defect structures can be modeled successfully by small supercell models. However, according to elasticity theory, the stress field of an isolated dislocation decays according to the factor of $1/r$ (Hirth and Lothe 1982) and, therefore, it is not possible to treat such an isolated dislocation within first principles calculations. Two approaches have been developed to evaluate the dislocation within the first principles framework. One is the flexible boundary condition method, which can capture the long-range strain field of a dislocation by coupling the dislocation core to an infinite harmonic bulk through the lattice Green function. The other is to consider the dislocation dipoles in a periodic cell. The elastic field of the dislocation dipole in a periodic cell can be treated by the superposition of an image dislocation (Cai et al. 2003, 2001) or by considering the elastic field in the reciprocal space (Mura 1964; Daw 2006). The latter is explained as follows.

The dislocation core in a periodic cell is introduced based on the solution of continuum linear elastic theory of the periodic dislocation dipole array (Daw 2006). The displacement gradient caused by the periodic distribution of dislocations is expressed as a Fourier series:

$$\underset{=}{\Delta}(\mathbf{r}) = \sum_{\mathbf{G}} \underset{=}{\tilde{\Delta}}(\mathbf{G}) \exp(i\mathbf{G} \cdot \mathbf{r}), \qquad (5.10)$$

where $\underset{=}{\Delta}(\mathbf{r})$ is the displacement gradient tensor at point $\mathbf{r}$, $\mathbf{G}$ is the reciprocal lattice vectors, and $\underset{=}{\tilde{\Delta}}(\mathbf{G})$ is the Fourier coefficients describing the displacement gradient tensor. The elastic energy per unit length of the supercell can be expressed in terms the $\underset{=}{\tilde{\Delta}}(\mathbf{G})$ s:

$$E_{\text{elastic}} = \frac{1}{2} A_c c_{ijkl} \sum_{\mathbf{G}} \tilde{\Delta}_{ij}(G) \, \tilde{\Delta}_{kl}^*(G), \qquad (5.11)$$

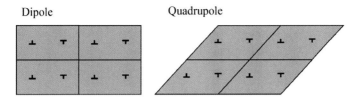

Fig. 5.12 Atomic model of dislocation dipole and quadrupole configurations in a periodic cell

Table 5.2 Lattice parameters and elastic constants in hexagonal Mg

	a (Å)	c/a	c_{11}	c_{33}	c_{12}	c_{13}	c_{44}
DFT	3.203	1.614	62.4	70.9	22.0	21.9	11.7
Exp. (Hearmon 1979)	3.209	1.623	59.3	61.5	25.7	21.4	16.4

where A_c is the area of the supercell perpendicular to the dislocation line and c_{ijkl} are the anisotropic elastic constants. The distortion field can be solved by minimizing the total elastic energy subject to the topological constraints imposed by the dislocation dipole. This theory is found in the literature, and the procedure is solved numerically by a sequence of linear algebra in the reciprocal space (Daw 2006). Subsequently, the displacement can be evaluated by a line integral starting from a reference coordinate in real space, and the local displacement at the atomic positions is provided. In practice, core radius, r_c is introduced to insure convergence of the summation for the elastic energy.

This approach is applied to HCP Mg alloys. The dipole of screw dislocation whose Burgers vectors is type $a/3\langle 11\bar{2}0 \rangle$ is treated here. The dislocation is inserted into a 288-atom supercell with 12×6 periodic units along [0001] and [$1\bar{1}00$], respectively, where the dislocation is inserted as dipolar and quadrupolar configurations, as shown in Fig. 5.12. The elastic constants and lattice parameters for Mg are obtained using density functional theory (DFT). Our computed values are shown in Table 5.2. The (2-D, for this problem) reciprocal lattice vectors can be expressed in terms of the primitive reciprocal lattice vectors, defined to be $\mathbf{a}^*$ and $\mathbf{b}^*$, as $\mathbf{G}_{hk} = h\,\mathbf{a}^* + k\,\mathbf{b}^*$ with $h, k \in Integers$. Seen in the predictions of one component of the displacement gradient tensor as a function of the number of $\mathbf{G}$ vectors in the sum, the Fourier series is completely converged when the core radius r_c is equal to b, and both h and k range from -20 to 20. In this way, the displacement gradient can be solved numerically. The solution of the strain and displacement fields of the dislocation dipolar and quadrupolar configurations evaluated by linear elasticity theory in a periodic cell are shown in Fig. 5.13. The strain field is actually periodic and the displacement field is depicted in the cumulative boundary condition along the y-direction. The method also allows us to estimate the elastic strain energy of dipolar and quadrupolar configurations in terms of the change in the relative displacement of the dislocation pair. The introduction of dislocation dipoles within each supercell results in a distortion of the supercell vectors, as noted by Lehto and Öberg (Lehto and

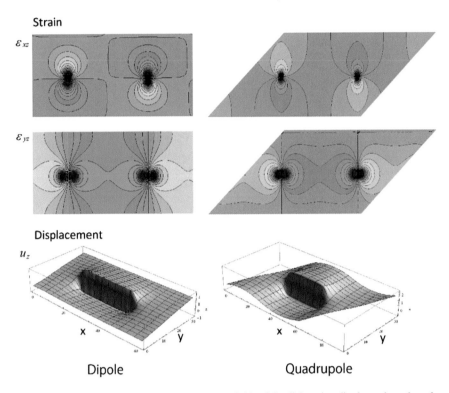

Fig. 5.13 Solution of the strain and displacement fields of the dislocation dipolar and quadrupolar configurations, evaluated by linear elasticity theory in a periodic cell

Öberg 1998). The relationship between strain energy and the relative displacement in the case of the dipolar and quadrupolar configurations in Mg with the above atomic model is given in Fig. 5.14. Here, the energy difference of the quadrupole is less sensitive to the relative displacement than that of the dipole, and therefore the quadrupolar configurations should be chosen for the first principles calculations when a small supercell is used.

5.4.2 First Principles Calculations

The displacements corresponding to each lattice point, obtained using the continuum linear elasticity theory outlined above, were applied to the atomic model for a dislocation dipole in the quadrupolar stacking configuration. In addition, the periodic supercell vectors were adjusted as discussed above. The dislocation core structure so obtained is used as the initial configuration for the first principles calculations. The supercell model of HCP metals is prepared as provided above, and the that of BCC metals is similarly prepared: a 135 atom model with dimensions $\mathbf{a} = 5\mathbf{e}_1$,

Fig. 5.14 Strain energy depends on the relative displacement when the size of a supercell is applied to the linear elasticity model

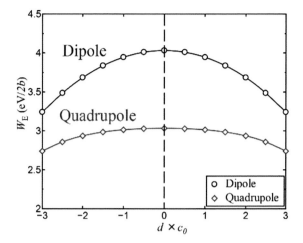

$\mathbf{b} = 2.5\mathbf{e}_1 + 4.5\mathbf{e}_2 + 0.5\mathbf{e}_3$, and $\mathbf{c} = \mathbf{e}_3$, where unit vectors in the system are defined as $\mathbf{e}_1 = a_0\left[\sqrt{6}, 0, 0\right]$, $\mathbf{e}_2 = a_0\left[0, \sqrt{2}, 0\right]$, and $\mathbf{e}_3 = a_0\left[0, 0, \sqrt{3}/2\right]$, with a coordinate system corresponding to the $x = [11\bar{2}]$, $y = [1\bar{1}0]$ and $z = [111]$ directions.

First principles calculations were then implemented to determine the structural relaxation. The stable configuration of the dislocation core structures in typical BCC and HCP metals is shown in Fig. 5.15, where the left side of the dislocation is depicted by the differential displacement (DD) vector (Vitek et al. 1970; Vitek 1974), and the core structures in BCC Fe, Nb, Mo and HCP Mg, Ti, Zr are provided as the typical examples. It is obvious from Fig. 5.15 that the dislocation core of BCC metals is compact owing to the high SF energy, as shown in Fig. 5.2. In addition, there is little difference in the dislocation core structures depending on the type of element in BCC metals. However, the core structures of HCP metals differ completely depending on type. As discussed in Sect. 5.2, dislocations of HCP metals tend to be dissociated according to the SF energy. The dislocation of Mg is dissociated in the basal plane due to the small SF energy in its plane while that of Zr dissociated in the prismatic plane. On the contrary, the dislocation of Ti is dissociated in the first-order pyramidal $(10\bar{1}1)$ plane. Pyramidal <a> dislocations cannot be activated because the SF energy along the <a> direction in the pyramidal $(10\bar{1}1)$ plane is generally extremely high. However, a unique core structure was recently discovered where the <a> dislocation is more likely to dissociate in the dense $(10\bar{1}1)$ plane than in prismatic plane because the SF energy, when the displacement is half the Burgers vector along the <a> direction in the dense $(10\bar{1}1)$ plane, is slightly smaller than that of the prismatic plane (Clouet et al. 2015). Thus, HCP metals have a wide variety of dislocation core structures, including various <a> dislocations and <c + a> dislocations in the first and second-order pyramidal planes, reflecting the SF energy, as shown in Fig. 5.5. As a result, plastic deformation in HCP metals becomes more complicated, i.e., solute atoms play a more significant role in the change in the slip mode through the electronic interactions between matrix elements and solutes.

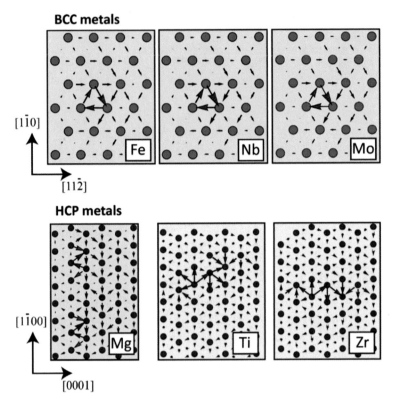

Fig. 5.15 Stable configurations of the dislocation core in BCC Fe, Nb, Mo, and HCP Mg, Ti, Zr, evaluated by first principles calculations. The relative displacement associated with the nearest neighbor atom was visualized by a DD vector

At the end of this section, the influence of solute atoms on the dislocation core structure and motion in Mg alloys is taken as an example. When the stable configuration was obtained, the energy barrier for the dislocation motion (i.e., Peierls potential) by first principles calculations attracts interest. The nudged elastic band (NEB) method (Henkelman et al. 2000) was employed to evaluate the electronic structure of the dislocation core when it overcomes the Peierls barrier. Initial configurations were prepared corresponding to the most stable structure position and the position where a dislocation moves through the Peierls barrier. The energy difference during dislocation glide of the left side of the dislocation on the basal, prismatic, and pyramidal planes, as well as the DD map at intermediated transition states, are shown in Fig. 5.16. It is confirmed that the screw dislocation dissociates into partial dislocations on the basal plane, which is caused by the nature of low SF energy of the basal plane in Mg. It is easy for this type of basal dislocation to glide on the basal plane—there is a negligibly small energy barrier between the symmetrically equivalent structures during basal slip. However, the prismatic slip requires surmounting high energy barrier by NEB calculation. The prismatic slip can be divided into three

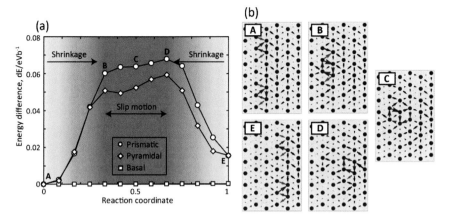

Fig. 5.16 Energy and core structure during the dislocation motion. **a** The minimum energy paths are for the motion of the screw dislocation in the basal, prismatic and pyramidal planes. **b** Intermediate transition states for prismatic slip visualized by the DD map (Tsuru and Chrzan 2015). Reprinted from Sci. Rep., vol. 5, Tsuru et al., Fig. 2

steps which is consistent with the Friedel–Escaig mechanism proposed by Couret and Caillard (Couret and Caillard 1985). At first, dissociated dislocation on the basal plane constricts that yields the compact core screw dislocation capable of gliding on the prismatic plane. Then, the dislocation with the compact core glides on the prismatic plane. Afterward, a compact core redissociates into partial dislocations on the basal plane. It is evident from Fig. 5.16 that a dominant part of the energy barrier for the prismatic slip is the energy for the constriction of two partial dislocations which enables to cross-slip.

This energy barrier is approximately 0.06 eV/b, with b being the Burgers vector of <a> dislocation. In contrast, the energy barrier for the glide of the compact core of screw dislocation is much lower, approximately 0.01 eV/b for both prismatic and pyramidal glide.

The energy barrier computed in this way has two general contributions: (I) the energy barrier for the motion of an isolated dislocation and (II) the changes to this barrier arising from the interaction with the periodic images. The contribution (II) needs to be considered rightly due to an artifact of the supercell geometry. As shown in Fig. 5.14., Daw's formulation of the periodic elasticity problem (Daw 2006) can be applied to estimate the changes in elastic energy within a periodic supercell as the dislocation moves between A and E. The linear elastic description of the periodic supercell enables to discern the effects of the periodicity on the computed Peierls barrier. According to elasticity theory, the contribution of the energy barrier as a function of supercell size can be drawn by Daw's formulation which found to be eliminated by considering in the large supercell limit (~18,000 atoms) (Tsuru and Chrzan 2015).

Fig. 5.17 Electronic structure of **a** the dissociated and **b** the compact dislocations in Mg. The projected DOS around the core is indicated by red and blue lines and that of the perfect crystal is indicated by a black line. Dotted lines and solid lines indicate s- and p-states, respectively. Energy states with respect to the energy range from -4.0 to 2.0 eV are shown in the inset (Tsuru and Chrzan 2015). Reprinted from Sci. Rep., vol. 5, Tsuru et al., Fig. 3

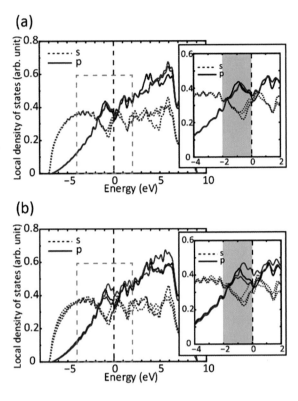

Returning to the quantum formulation of the problem, the changes in the electronic structure associated with the dislocation motion along the prismatic plane. The electronic structures of the initial and the sixth intermediate configurations corresponding to the partial and compact dislocations, are shown in Fig. 5.17. The site-projected density of states (DOS) on the two atom positions around the dislocation core for the dissociated and compact cores corresponding to states A and C in Fig. 5.16 was examined as well as the atom in the perfect crystal. Thus, the evolution of the site-projected DOS as the dislocation passes over the Peierls barrier can be captured.

As presented in Fig. 5.17, analyzing the DOS establishes two points. First, the states of the outermost s electrons of Mg do not change significantly in the dislocation core. Second, the passage of a dislocation is influenced strongly by the p-states just below the Fermi level. The DOS around a site of core region is reduced by the compact dislocation while the projected DOS for the other site is increased, which means the DOS changes slightly when the dislocation forms unstable compact core. Even if the change of DOS is not so significant, the rearrangements are much more pronounced in the case of the compact core. Furthermore, these p-states participate in the bonding rearrangements that take place as the core transitions from partials to a compact core to surmount its Peierls barrier. The changes of the electronic structure as it surmounts its Peierls barrier constitute the electronic structure during the dislocation motion.

It is thus interesting to identify substitutional elements which is expected to interact significantly with the dislocations.

The principle underlying our analysis is not so complicated. Just see how the electronic structure changes when a substitutional solute is introduced around a dislocation core. It is expected that a strong chemical hybridization can exist between electronic states of solute atom and Mg atoms around solute and dislocation core if the changes occur in the same energy range associated with them. To explore this idea, the electronic structures of Al, Zn, Y, Ca, Ti, and Zr substitutional atoms within Mg were examined. Figure 5.18 shows the site-projected DOS of various solutes around the core. Here, Al and Zn have small effect on the electronic states in the energy range (-2–0 eV) which is important for the motion of the dislocation. In contrast, Y, Ca, Ti, and Zr are found to have relatively large changes in this energy range. Therefore, it is expected that substitutional Y, Ca, Ti, and Zr have strong interaction with the dislocation motion while Al and Zn will not.

The interaction of the dislocation with substitutional solute atoms can be considered directly based on this analysis. Figure 5.19 provides a typical comparison between solutes. Even if the Al solute is added around core region, the dislocation remains dissociated partial dislocations as the Al solute does not influence the

Fig. 5.18 Electronic structure of Mg and a solute in a defect-free crystal. **a** Projected DOS for three atoms around Al and Y solutes. Curves for the perfect region are shown with black lines. **b** DOS for solute elements (Al, Zn, Y, Ca, Ti, and Zr) in a defect-free crystal (Tsuru and Chrzan 2015). Reprinted from Sci. Rep., vol. 5, Tsuru et al., Fig. 4

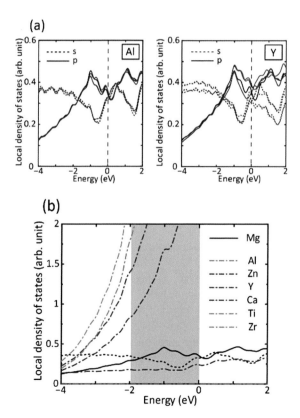

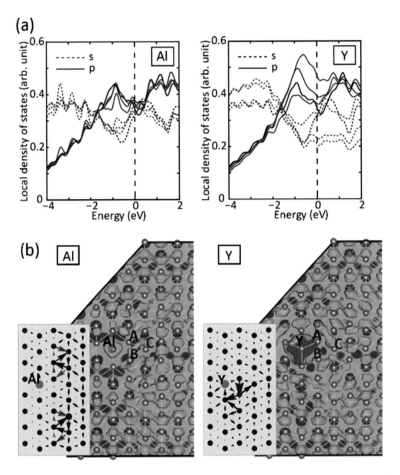

Fig. 5.19 Local electronic structure of the dislocation core around a solute. **a** Projected DOS for atoms constituting the dislocation core. **b** DD map and partial charge density for the dislocation core with a solution with respect to a specific energy range (−2.0–0 eV) (Tsuru and Chrzan 2015). Reprinted from Sci. Rep., vol. 5, Tsuru et al., Fig. 5

electronic states near the dislocation core. In contrast, due to Y solute which have a strong interaction make a dissociated dislocation in the basal plane shrink immediately. Of note, the partial charge density for a dislocation core with Al and Y shows a similar distribution to those for a dissociated and compact dislocation cores, respectively, in pure Mg. Figure 5.19a shows the DOS of Al and Y solutes, where the *s*- and *p*-states of Mg and the *d*-states of each solute element are given in the same graph.

There is clear difference between two types of elements: Al and Zn have no or few *d*-states in the energy range of interest. In contrast, Y, Ca, Ti, and Zr show prominent *d*-states below the Fermi level. These *d*-states are possible for hybridization with the *p*-states of the Mg, and ultimately lead to a strong interaction between these alloying additions and the dislocation core structure. These calculations, therefore, confirm

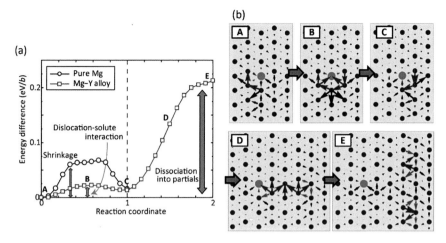

Fig. 5.20 Effect of a solid solution strengthening and softening. **a** Energy differences during the dislocation motion in the prismatic plane in pure Mg and an Mg–Y. The energy curve when the dislocation passes through the Y solution is indicated by the red line. **b** Core structure during the dislocation overcoming and moving away from the Y solution (Tsuru and Chrzan 2015). Reprinted from Sci. Rep., vol. 5, Tsuru et al., Fig. 6

that the electronic structure of the dislocation motion can be applied to identify alloying elements that interact strongly with the dislocations in Mg.

In pure Mg, the glide of a screw dislocation in the prismatic plane is only possible when the dislocation has the compact core configuration as described above. According to the electronic structure calculations show, solutes with the strong interaction bind to, and thus stabilize the compact core. To confirm this scenario, the NEB calculation was reemployed, in which a column of Mg atoms was replaced by Y atoms and the results are shown in Fig. 5.20. Figure 5.20 provides two important observations. First, Y solute results in the core transformation from dissociate partials to compact core when Y is located around dislocation core region. Second, the influence of Y solute is relatively wide-ranged, approximately 3/2 c in all directions. Third, the interaction between Y and dislocation is quite significant, which increase the energy barrier for the basal slip.

Again, plastic anisotropy is one origin of the poor ductility in Mg. In the absence of solute atoms, screw dislocations glide frequently on the basal plane, whereas the energy barrier for non-basal slip are quite high. This plastic anisotropy can be improved if the strength of basal plane slip increases and non-basal slip becomes more frequent. Some solutes found to have a great potential to increase the barrier for basal slip as decreasing the barrier of non-basal slip. It becomes more important to capture the chemical interaction between the dislocation and solutes from electronic structure basis (Tsuru 2017).

In summary, dislocation motion has been recognized as one of the most important plastons to determine macroscopic plastic deformation. In this section, several approaches were introduced to capture the dislocation core structure and dislocation-related properties through first principles calculations. These approaches have great potential in the understanding of the fundamental role of dislocation in plastic deformation grounded in electronic structure calculations.

Acknowledgements T.T acknowledges the support of JST PRESTO Grant Number JPMJPR1998 and JSPS KAKENHI (Grant Numbers JP19K04993, JP18H05453). Simulations were performed on the large-scale parallel computer system with SGI ICE X at JAEA.

References

Bulatov VV (1997) Phys Rev Lett 78:4221
Cai W, Bulatov VV, Chang J, Li J, Yip S (2001) Phys Rev Lett 86:5727
Cai W, Bulatov VV, Chang J, Li J, Yip S (2003) Philos Mag 83:539
Carter CB, Holmes SM (1977) Philos Mag 35:1
Clouet E, Caillard D, Chaari N, Onimus F, Rodney D (2015) Nature Mater 14:931
Couret A, Caillard D (1985) Acta Metall 33:1447
Daw MS (2006) Comput Mater Sci 38:293
Dobson P, Goodhew P, Smallman R (1967) Phil Mag 16:9
Ercolessi F, Adams JB (1994) Europhys Lett 26:583
Finnis MW, Sinclair JE (1984) Philos Mag A 50:45
Foreman AJE (1955) Acta Metall 3:322
Hearmon RF (1979) The elastic constants of crystals and other anisotropic materials. In: Hellwege K-H, Hellwege AM (eds) Landolt-Börnstein Tables, Group III, vol 11. Springer, erlin
Henkelman G, Uberuaga BP, Jónsson H (2000) J Chem Phys 113:9901
Hirth JP, Lothe J (1982) Theory of dislocations, 2nd edn. Wiley, New York
Joós B, Duesbery MS (1997) Phys Rev Lett 78:266
Kresse G, Furthmuller J (1996) Phys Rev B 54:11169
Kresse G, Hafner J (1993) Phys Rev B 47:558
Lehto N, Öberg S (1998) Phys Rev Lett 80:5568
Li J (2003) Model Simul Mater Sci Eng 11:173
Liu XY, Adams JB, Ercolessi F, Moriarty JA (1996) Model Simul Mater Sci Eng 4:293
Lu G, Kioussis N, Bulatov VV, Kaxiras E (2000) Phys Rev B 62:3309
Mishin Y, Farkas D, Mehl MJ, Papaconstantopoulos DA (1999) Phys Rev B 59:3393
Mishin Y, Mehl MJ, Papaconstantopoulos DA (2005) Acta Mater 53:4029
Mishin Y, Mehl JJ, Papaconstantopoulos DA, Voter AF, Kress JD (2001) Phys Rev B 63:224106
Monkhorst HJ, Pack JD (1976) Phys Rev B 13:5188
Mura T (1964) Proc Roy Soc A 280:528
Murr L (1975) Interfacial phenomena in metals and alloys. Addison-Wesley, India
Nabarro FRN (1947) Proc Phys Soc 59:256
Orowan E (1934) Z. Physik 89:605
Peierls R (1940) Proc Phys Soc 52:34
Perdew JP et al (1992) Phys Rev B 46:6671
Polanyi M (1934) Z. Physik 89:660
Sun DY, Mendelev MI, Becker CA, Kudin K, Haxhimali T, Asta M, Hoyt JJ, Karma A, Srolovitz DJ (2006) Phys Rev B 73:024116
Taylor GI (1934) Proc Roy Soc A 362
Tsuru T (2017) Bull Jpn Inst Metals Materia 56:5 (in Japanese)
Tsuru T, Chrzan DC (2015) Sci Rep 5:8793

Tsuru T, Udagawa Y, Yamaguchi M, Itakura M, Kaburaki H, Kaji Y (2013) J Phys: Condens Matter 25:022202
Vítek V (1966) Phys Status Sol 18:687
Vítek V (1968) Philos Mag 18:773
Vitek V (1974) Cryst Latt Def 5:1
Vitek V, Perrin RC, Bowen DK (1970) Philos Mag 21:1049
Volterra V (1907) Ecol Norm Sup 324:405

Part III
Experimental Analyses of *Plaston*

Chapter 6
Plaston—Elemental Deformation Process Involving Cooperative Atom Motion

Haruyuki Inui and Kyosuke Kishida

6.1 Introduction

There is an ever-increasing demand for structural materials that simultaneously possess high strength and high ductility/toughness. It is, however, very difficult to achieve this in any material because strength and ductility are in general in the trade-off relationship. The material of high strength generally exhibits low ductility/toughness (as in many ceramics) and vice versa. Plastic deformation of crystalline materials usually occurs by shear deformation along a particular crystallographic plane (referred to as slip) so that the shearing of one portion of a crystal occurs with respect to another (by the vector called 'slip vector') upon the crystallographic plane (called 'slip plane'), and such slip is usually (in many cases) carried by a lattice defect called 'dislocation', which is defined as the boundary (line defect) between slipped and unslipped regions on the crystallographic plane (slip plane) (Fig. 6.1). The dislocation can move on the slip plane under the exerted shear stress, displacing one portion of a crystal with respect to another by the vector called 'Burgers vector'. If the stress required to move dislocations is high, the strength of the material is high but the ductility is low because of the difficulty in the dislocation motion. If the stress required to move dislocations is low, in contrast, the strength of the material is low but the ductility is high due to the ease in the dislocation motion.

'Dislocation' is a line defect characterized by the Burgers vector and line vector. By definition, all atoms of one portion of a crystal along the dislocation line are

H. Inui (✉) · K. Kishida
Department of Materials Science and Engineering, Kyoto University, Sakyo-ku, Kyoto 606-8501, Japan
e-mail: inui.haruyuki.3z@kyoto-u.ac.jp

K. Kishida
e-mail: kishida.kyosuke.6w@kyoto-u.ac.jp

Center for Elements Strategy Initiative for Structural Materials (ESISM), Kyoto University, Sakyo-ku, Kyoto 606-8501, Japan

© The Author(s) 2022

I. Tanaka et al. (eds.), *The Plaston Concept*,
https://doi.org/10.1007/978-981-16-7715-1_6

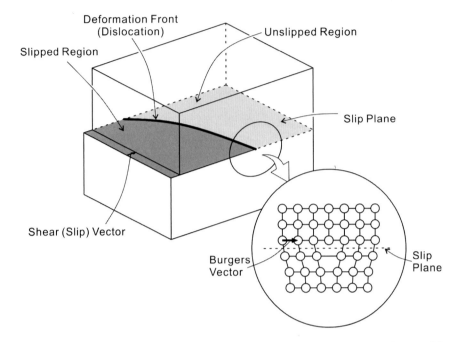

Fig. 6.1 General description of the elemental process of plastic deformation of crystalline materials
with the passage of the deformation front called 'dislocation'

displaced 'uniformly' by an amount determined by the Burgers vector upon the
dislocation motion so that the region of the atom displacement occurring in a line
(that is dislocation) propagates on the slip plane in the crystal under the exerted
shear stress to complete shear (plastic) deformation (Anderson et al. 2017; Nabarro
1967; Friedel 1964; Weertman and Weertman 1964). In this sense, the dislocation
is the deformation front that makes cooperative atom displacement, the magnitude
of which is determined by the slip (Burgers) vector. One point we should notice
here is that the crystal adopts the lowest energy process in plastic deformation, and
it is the motion of dislocations that is the lowest energy process in many cases in
crystalline materials. The ease of such cooperative atom displacement (along the
dislocation line) basically determines the stress necessary to move the dislocation
and its mobility, and hence, is closely related to the strength and ductility/toughness
of the material. Of course, the strength and ductility/toughness are not determined
solely by the intrinsic nature of the crystalline lattice of the material as described
above and are dependent also on microstructural features such as grain size and
incorporation of alloying elements and other phases (Anderson et al. 2017; Nabarro
1967; Friedel 1964; Weertman and Weertman 1964). The material property has thus
been altered in positive ways by tuning these microstructural parameters. One way
to break the trade-off relationship between strength and ductility/toughness is to
activate a deformation mode different from the dislocation discussed above. Since

Fig. 6.2 Schematic
stress-strain curve for
materials with increased
strength and ductility by
subsequent activation of
different deformation modes

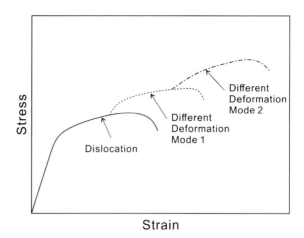

the dislocation is a simple lattice defect that can move relatively easily, the initial plastic flow would be carried by these dislocations. But, if a deformation mode different from 'dislocation' can operate once these dislocations encounter difficulty in their further motion in the course of deformation, both strength and ductility are expected to increase simultaneously (Fig. 6.2). Deformation modes different from the dislocation include dislocations with a different Burgers vector (such as those with the **a+c** Burgers vector in HCP crystals) and those with a different shear mechanism (such as synchroshear (Kronberg 1957) and zonal shear (Kronberg 1959, 1961)), twinning (Read-Hill et al. 1963; Cristian and Mahajan 1995), phase transformation and something else (related to grain boundary motion and so on) (Anderson et al. 2017; Nabarro 1967; Friedel 1964; Weertman and Weertman 1964). TWIP and TRIP steels (Frommeyer et al. 2003; Cooman et al. 2018) as well as, more recently, high-entropy alloys with the FCC structure (Otto et al. 2013; Gludovatz et al. 2014) are some nice examples, in which simultaneous achievement of high strength and high ductility is realized by inducing the activation of different deformation modes following the starvation of the primary mode of dislocation.

It seems very important to understand how we can activate not only the primary deformation mode (the simple dislocation) but also the secondary (and tertiary) deformation mode(s), in terms of stress level/state, crystallographic orientation, grain size and orientation relationships with neighboring grains and so on for the development of structural materials with high strength and ductility/toughness. We call all these deformation modes including the primary dislocation 'plaston' (see the reason below) and discuss the importance of the understanding of the activation of various plastons for achieving high strength and ductility/toughness.

6.2 Nucleation and Motion of Plastons (Possible Deformation Modes) Under Stress

Crystalline materials generally contain grown-in dislocations that are introduced during solidification (and forming as well) unless otherwise some particular care is taken. Grown-in dislocations are in general simple dislocations that can move at the lowest stress levels (among the available deformation modes in the material), since they are introduced most easily in order to relieve internal stresses developed during solidification. If the stress level for the dislocation motion is too high (as in many brittle materials), the material cracks instead of introducing dislocations. Upon plastic deformation, these grown-in dislocations move and multiply new dislocations once external stress sufficiently high enough to move them is applied to the material. This is because these dislocations are the weakest parts of the material under stress such that cooperative atom displacements along the dislocation line can occur most easily, when compared to otherwise perfect portions without any lattice defects. Multiplication of dislocations from these grown-in dislocations is what usually occurs in most bulk crystalline materials during deformation. As the specimen size becomes smaller down to sub-micron meter sizes, however, the possibility to find out grown-in dislocations in the specimen becomes lower, such that plastic deformation occurs by the nucleation of new dislocations and their subsequent motion (but not by multiplication of dislocations from the pre-existing grown-in dislocations), as have frequently been seen in whisker and micro- and nano-pillar specimens (Uchic et al. 2004; Okamoto et al. 2013, 2014, 2016; Chen et al. 2016; Zhang et al. 2017; Kishida et al. 2018, 2020; Higashi et al. 2018). When there is no available pre-existing (grown-in) dislocation at the beginning of plastic deformation, the stress level for deformation is known to be quite high from experiments with whisker and micro- and nano-pillar specimens.

One of the typical examples that are accompanied by the nucleation of dislocation during plastic deformation is shown in Fig. 6.3 for the case of micropillar compression tests made for two different orientations of single crystals of Ti$_3$Al with the DO$_{19}$ (ordered HCP) structure. Because of the brittleness of this particular intermetallic together with the small specimen size, it is fairly a safe assumption that there are virtually no grown-in dislocations available at the beginning of plastic deformation. Two different deformation modes are activated depending on crystal orientation. Pyramidal **a**+**c** ($\{11\bar{2}1\}<11\bar{2}6>$) slip with the highest CRSS value in the bulk is observed to operate in the whole specimen size range investigated for the [0001] orientation, while pyramidal **a**+**c** ($\{11\bar{2}1\}<11\bar{2}6>$) slip is replaced by prism **a** ($\{1\bar{1}00\}<11\bar{2}0>$) slip as the specimen size decreases for the [$2\bar{1}\bar{1}0$] orientation with the slip system transition occurring at around 6 μm. In view of the fact that grown-in dislocations, if exist, are expected to have a Burgers vector **b** = 1/3<11$\bar{2}$0>, plastic deformation is believed to occur by the nucleation of dislocations and their subsequent motion. This is believed to be the case not only for the [0001] and [$2\bar{1}\bar{1}0$] oriented micropillar specimens where **a**+**c** dislocations are activated but also for smaller micropillar specimens with the [$2\bar{1}\bar{1}0$] orientation, since the CRSS (critical resolved shear stress) values for prism **a** slip obtained in micropillars are considerably

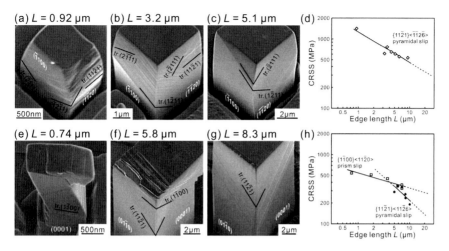

Fig. 6.3 Micropillar compression tests on **a-d** [0001]- and **e-h** [2$\overline{1}$$\overline{1}$0]-oriented Ti₃Al single crystals

higher than that for the bulk. If we assume that the yield stress (CRSS) observed in micropillar compression tests is well correlated with the nucleation stress for dislocations, the nucleation stress for dislocations of a particular slip system obviously varies with the specimen size. The fact that the operative deformation mode changes with the specimen size even with the same orientation (the slip system transition for the [2$\overline{1}$$\overline{1}$0] orientation in Ti₃Al) clearly indicates that the selection of operative deformation mode is the subject that is determined by stress level/state, crystallographic orientation, specimen size (grain size) and so on. Although the above example deals with dislocations with different Burgers vectors (**a** and **a+c**), it is obvious that other deformation modes including twinning, phase transformation and something else (related to grain boundary motion and so on) can all be subject to be considered for the nucleation under stress as deformation modes to carry plastic strain, as will be detailed in the next section.

When deformation is made along a fixed path (as in uniaxial tensile and compression tests), lattice instability starts to occur as the stress (strain) increases so that the saddle (bifurcation) point is passed (Fig. 6.4). The lattice instability may be evaluated, for example, by the positivity of elastic stiffness constants (Born's lattice instability criteria) (Yashiro 2012). If this occurs in some local area of the specimen (as can happen due to stress concentration at (even atomic) steps on the specimen surface), the area of such lattice instability is considered to contain atoms that are cooperatively exited and that change their arrangement energetically suitable to the exerted stress/strain state (cooperative atom displacement). Such a change in the atomic arrangement under stress produces a dislocation in some cases and other deformation carriers (twinning, phase transformation and so on) in other cases. We believe this is the nucleation of plastons (Fig. 6.5a). The nucleation of plastons may occur at steps on the specimen surface (more precisely, on the corner edge of the specimen) to avoid any significant constraint from the surroundings in the case of

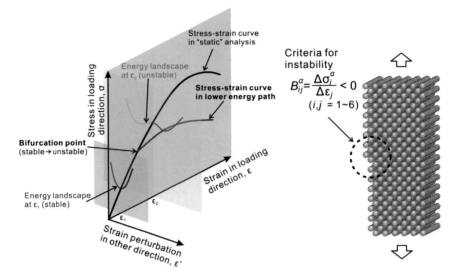

Fig. 6.4 Lattice instability occurring during deformation along a fixed path at high stress levels, where the positivity of elastic stiffness constants is violated (after Yashiro (2012))

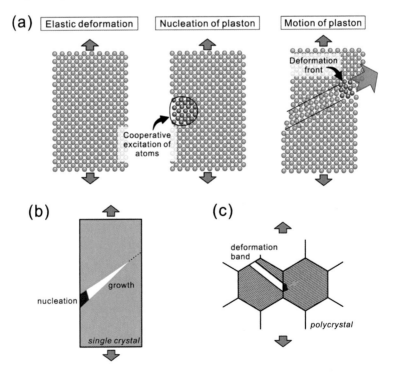

Fig. 6.5 a General schematics of nucleation and motion of plaston. Nucleation of plaston in **b** single crystal and **c** polycrystals

single-crystalline micro- and nano-pillars (Fig. 6.5b). But, the nucleation of plastons at the grain boundary must be more important when considering bulk polycrystalline materials in practical use for deformation propagation across the grain boundary (Fig. 6.5c). We also believe that the deformation mode (plaston) selected among possible deformation modes must be that which involves the atomic process of the lowest energy depending on stress level/state, crystallographic orientation, specimen size (grain size) and so on for a particular material. Their subsequent motion must also be dependent on stress level/state, crystallographic orientation, specimen size (grain size) and so on, so that the deformation mode difficult to move may not be selected to contribute to the plastic flow of the material. Atoms that are cooperatively excited are generated locally under stress and they change the atomic arrangement to energetically fit with the exerted stress/strain state, crystallographic orientation, orientation relationship with neighboring grains (in the case of polycrystals) and so on during deformation, generating (nucleating) the deformation carries (plastons), and they spread as a line defect (in many cases, along a particular crystallographic plane) in the crystal to carry plastic strain. We call these deformation carriers 'plastons'. We have to think of both the nucleation and motion of these plastons, as in the case of dislocations. Needless to say, these deformation carriers are not necessarily dislocations but can be others.

6.3   Cooperative Motion of Atoms in Plastons

When grown-in dislocations are available, they move and multiply new dislocations once external stress sufficiently high enough to move them is applied to the material (Anderson et al. 2017; Nabarro 1967; Friedel 1964; Weertman and Weertman 1964). In this case, these dislocations are the weakest parts of the material under stress such that cooperative atom displacements along the dislocation line can occur most easily. In this way, simple dislocations are plastons (deformation carriers) that carry plastic flow at least in the early stages of plastic deformation in many conventional metallic materials (the inset of Fig. 6.1).

When these grown-in dislocations are not available (as in nano- and micro-pillars, ultrafine-grained materials and hard and brittle materials), the nucleation of plastons (deformation carriers) is needed to initiate plastic flow so that atoms that are cooperatively exited change their arrangement energetically suitable to the exerted stress/strain state (cooperative atom displacement). The simplest way to achieve this may be to nucleate simple dislocations. But, if this is not possible (for example, by the reason of the unfavorable crystal orientation, unfavorable orientation relationship with neighboring grains, too high stress level for simple dislocations due to work-hardening and so on), plastons (deformation carriers) different from simple dislocations will be selected at stress levels higher than that for simple dislocations. These plastons nucleated at higher stress levels include dislocations with a different Burgers vector (such as those with the **a+c** Burgers vector in HCP crystals) and those with a different shear mechanism (such as synchroshear (Kronberg 1957) and zonal shear (Kronberg 1959, 1961)), twinning (Read-Hill et al. 1963; Cristian and Mahajan 1995), phase transformation and something else (related to grain boundary motion and so on)) (Anderson et al. 2017; Nabarro 1967; Friedel 1964; Weertman and Weertman 1964).

Here, we take an example of twinning as a plaston to explain cooperative atom displacement occurring in the nucleation and in the propagation as the deformation front during plastic deformation. In deformation twinning in HCP metals, the nucleation of a twin is known usually to occur at the surface or at the grain boundary. The nucleation of a twin may occur in some area that contains atoms cooperatively exited and change their arrangement energetically suitable to the exerted stress/strain state, crystallographic orientation, specimen size (grain size) and so on for a particular material. This is discussed in detail by Tanaka and Togo in a different chapter of this book and will be briefly described in the next section. Then, the area propagates as a plaston (deformation front) making a complicated cooperative atomic motion as frequently called 'shuffling', as this cannot be expressed with a single displacement vector (Fig. 6.6b). This occurs because HCP metals are 'diatomic' having two atoms at each of the primitive hexagonal lattice points, in contrast to the case of FCC and BCC metals, in which only a single atom is allocated to each of the face-centered and body-centered cubic lattice points ('monoatomic'), respectively. Displacements of atoms in the plaston (deformation front) are thus the same for all atoms in FCC and BCC metals, coinciding exactly to the Burgers vector of the 'twinning' partial dislocation. However, displacements of atoms in the plaston (deformation front) in 'multi-atomic' materials (with multiple atoms at each lattice point) generally differ from atom to atom and they do not generally coincide with the displacement vector of the relevant lattice points. In this case, the displacement vector of the relevant lattice

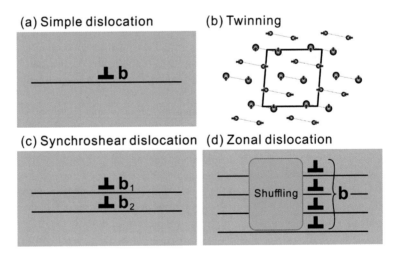

Fig. 6.6 Plastons of different types; **a** simple dislocation, **b** twinning, **c** synchroshear dislocation and **d** zonal dislocation. Compare the simple dislocation in the inset of Fig. 6.1

points is often assigned as the Burgers vector of the twinning dislocation for convenience. In this sense, the 'twinning dislocation' in 'multi-atomic' materials is quite different from that in 'monoatomic' materials. The fact that all atom displacements cannot be described simply with the Burgers vector indicates clearly that propagation of the plaston (deformation front) indeed includes the cooperative motion of atoms of the lowest energy under stress.

There are some particular dislocations that may involve a volume for cooperative atom motion probably larger than that for simple dislocations during their nucleation and motion. The way of cooperative atom motion for these dislocations is necessarily more complicated because the shearing process with these dislocations involves a few to several successive parallel atomic planes, in contrast to simple dislocations for which the shearing process involves a single atomic plane. The 'synchroshear' dislocation involves multiple (two, in many cases) successive parallel slip planes displaced synchronously along different (Burgers) vectors on each of the planes so as to give the displacement corresponding to the sum of the all vectors as a whole (Fig. 6.6c). This concept was first predicted for basal slip in alumina (Kronberg 1957), then for slip in compounds of the Laves phase (Chu and Pope 1984; Hazzledine and Pirouz 1993). The concept was proved experimentally first in transition-metal silicides with the C40 structure (Moriwaki et al. 1997; Inui et al. 1998) and then for basal slip in Laves-phase compounds (Kumar and Hazzledine 2004). The 'zonal' dislocation involves much more parallel slip planes called 'shear zone' in which atoms in the slab move non-uniformly but cooperatively by varying amounts (Fig. 6.6d). A typical example is prism **a+c** slip in HCP metals, as detailed in the past (Kronberg 1961). Slip in sigma-phase compounds is another example, in which atomic rings in the Kagome layers are forced to make a rotation in a cooperative manner. This was proved experimentally to occur for {110}<001> slip recently (Kishida et al. 2020).

We believe that the nucleation and motion of plastons are important also in bulk metallic glasses [for example, (Kumar et al. 2010)]. In bulk metallic glasses, plastic deformation is known to occur by the passage of the so-called shear zone (Ogata et al. 2006). Cooperative atom motion must occur at the front of the shear zone to energetically fit the exerted stress, sometimes generating another glass structure of the lower energy, referred to as rejuvenation (Meng et al. 2012).

All these indicate that the cooperative and complicated atom motion occurs in general in the deformation front under stress in many different materials. Of importance to remember is that such cooperative and complicated atom motion is sometimes energetically favored under certain circumstances in particular materials. It is thus very important to consider how we can control the occurrence of such cooperative and complicated atom motion.

6.4　Origin of Cooperative Atom Motion in the Nucleation of Plastons

Plastons that make an energetically favorable cooperative atom motion will be nucleated depending on the exerted stress/strain state, crystallographic orientation, orientation relationship with the neighboring grain and specimen size (grain size), as discussed above. We now show how the selection of deformation mode (carrier) would be made by taking an example of deformation twinning in HCP metals only briefly, since this is discussed in detail by Tanaka and Togo (Tanaka and Togo this book) in a different chapter of this book. As discussed in Sect. 6.2 of this paper, the nucleation of plastons will be made in some local area of the specimen in which lattice instability occurs as a result of high stress/strain. Then, the question is how such lattice instability occurs. Tanaka and Togo have indicated that phonon instability occurring as the shear stress/strain increases triggers the cooperative atom rearrangement accompanied by crystal symmetry breaking, resulting in $\{10\bar{1}2\}$ twinning in Ti in their first-principles calculations. According to them, the cooperative atom motions during shuffling are correctly predicted. The selection of $\{10\bar{1}2\}$ twinning as the plaston to carry plastic strain occurs of course only when the crystal is oriented favorably for its occurrence. Otherwise, the selection of other modes will be energetically more favored. Although yet to be proved by first-principles calculations, slip with dislocations of the particular type will be similarly selected as the lowest energy process driven by lattice instability depending on crystal orientation (the magnitude of shear stress acting on the relevant deformation modes). Of importance to note is that the selection of plaston is made as the lowest energy process driven by lattice instability that triggers the cooperative atom rearrangement, depending on the exerted stress/strain state, crystallographic orientation, orientation relationship with the neighboring grain and specimen size (grain size).

6.5　Applications of the Concept of Plastons to the Improvement of Mechanical Properties of Structural Materials

Structural materials are used usually in the form of polycrystals mostly for economic reasons, except for some examples such as Ni-based superalloys used as turbine blades. This means that the concept of plastons should be applied effectively to polycrystalline materials in practical use for the improvement of mechanical properties of existing structural materials and for the development of new materials with high strength and high ductility/toughness. Although plastons are explained and discussed mainly with single crystals for ease of understanding in this paper, the nucleation of plastons at the grain boundary has to be taken into account (Fig. 6.5c). When deformation propagates across the grain boundary in polycrystalline materials, the nucleation of deformation mode (carrier) occurs in a new grain adjacent to the already-deformed grain or at the grain boundary between the two grains so as to satisfy strain compatibility across the grain boundary. Which deformation mode is selected upon deformation propagation across the grain boundary would be determined by the stress level/state and crystallographic orientation of the new grain (magnitude of shear stress), orientation relationship between the two adjacent grains (strain incompatibility), stress level/state of the grain, grain size and so on. Generally speaking, if the grain size is large on the milli-meter order, the deformation mode with the lowest CRSS (simple dislocations in most cases) is usually selected, unless there is some special orientation relationship among grains. In this case, slip carried by simple dislocations is observed in all grains throughout the specimen. This is actually the main cause of the low formability of many HCP metals, in which deformation modes with the **c**-axis component are hardly activated because of the higher CRSS values. However, there is an increasing possibility to activate deformation modes other than simple dislocations as the stress level increases, for example, by reducing the grain size (ultra fine-grained materials) and by strengthening through solid-solution hardening (high-entropy alloys). If some different deformation modes are activated as the stress level increases, following the primary slip with simple dislocations, high strength and high tensile ductility are expected to simultaneously be achieved (Fig. 6.2).

TWIP and TRIP steels (Frommeyer et al. 2003; Cooman et al. 2018) and high-entropy alloys (Otto et al. 2013; Gludovatz et al. 2014) are a good example for this. In addition to slip by simple dislocations, partial dislocations to produce twins and/or ε-martensite are activated at high stress/strain levels, leading to high strength and ductility. Obviously, the stacking fault energy and the relative energy difference between the relevant FCC and HCP phases are parameters to control the activation of partial dislocations to produce twin and/or ε-martensite, as the twinning stress, for example, has been believed to correlate positively with the stacking fault energy (Venables 1961). Of interest to note is that the opposite trend (negative correlation between the twinning stress and stacking fault energy) is recently reported (Laplanche et al. 2017), indicating that there is still enough room to investigate how we can activate partial dislocations to produce twins and/or ε-martensite in FCC-based alloys.

Ultra fine-grained materials may be another good example of this. As the grain size decreases, the strength of these ultra fine-grained materials is increased, but at the same time, the ductility generally decreases catastrophically. However, once deformation modes other than slip by simple dislocations are activated, the improvement of ductility is sometimes observed (Tsuji et al. 2020). The activation of partial dislocations to produce twins and/or ε-martensite, other dislocations with different Burgers vectors, disclinations and grain boundary sliding may contribute to the improvement of ductility of ultra fine-grained materials. An example is shown for ultra fine-grained Mg with the HCP structure in Fig. 2 of (Tsuji et al. 2020). Pyramidal **a+c** slip is observed to operate, in addition to primary basal **a** slip, once the grain size is reduced below $1\sim2$ μm, significantly improving the tensile ductility. This is an example showing how a high-stress state achieved by reducing the grain size can activate a deformation mode other than simple dislocations. The change in the *c/a* axial ratio by alloying is known to be effective to induce deformation modes with the axis component in HCP metals (for example, Al additions to Ti, and Zn and rare-earth element additions to Mg).

However, almost nothing is known about how we can effectively activate these additional deformation modes at high stress/strain levels in a general sense. Although some examples are shown above, the effects of orientation relationships among adjacent grains (texture) including those with the hard secondary phase grains have remained largely unknown in spite of the primary importance for actual structural materials in practical applications. It seems very important to investigate how we can effectively activate these additional deformation modes with the concept of plastons as outlined in this paper for the improvement of mechanical properties of existing structural materials and the development of new materials with high strength and high ductility.

6.6 Conclusions

Plaston is defined as a defect that is nucleated and propagates as a line defect of the deformation front to carry plastic flow under shear stress. The nucleation of plastons occurs in some local area of the specimen of lattice instability such that atoms in that local area are cooperatively exited and change their arrangement energetically suitable to the exerted stress/strain state (cooperative atom displacement) to form a plaston of the lowest energy. The selection of a plaston of a particular type among many different plastons depends on stress level/state, crystallographic orientation, specimen size (grain size) and so on. The importance of the understanding of the activation of various plastons is discussed.

Acknowledgements This work was supported by the Elements Strategy Initiative for Structural Materials (ESISM) of MEXT (Grant number JPMXP0112101000), in part by JSPS KAKENHI (Grant numbers JP18H01735, JP18H05478, JP18H05450, JP18H05451, JP19H00824, JP19K22053 and JP20K21084), and in part by JST CREST (Grant number JPMJCR1994).

References

Anderson PM, Hirth JP, Lothe J (2017) Theory of dislocations, 3rd edn. Cambridge University Press, Cambridge
Chu F, Pope DP (1984) Twinning on advanced materials (Yoo MH, Wuttig M (eds)). TMS, Warrendale, p 415
Cristian JW, Mahajan S (1995) Prog Mater Sci 39:1
Chen ZMT, Okamoto NL, Demura M, Inui H (2016) Scripta Mater 121:28
De Cooman BC, Estrinand Y, Kim SK (2018) Acta Mater 142:283
Friedel J (1964) Dislocations. Pergamon, Oxford

Frommeyer G, Brux U, Neumann P (2003) ISIJ Int 43:438
Gludovatz B, Hohenwarter A, Catoor D, Chang EH, George EP, Ritchie RO (2014) Science 345:1153
Hazzledine PM, Pirouz P (1993) Scripta Metall 28:1277
Higashi M, Momono S, Kishida K, Okamoto NL, Inui H (2018) Acta Mater 161:161
Inui H, Moriwaki M, Ito K, Yamaguchi M (1998) Phil Mag A 77:375
Kishida K, Maruyama T, Matsunoshita H, Fukuyama T, Inui H (2018) Acta Mater 159:416
Kishida K, Shinkai Y, Inui H (2020) Acta Mater 187:19
Kishida K, Okutani S, Inui H (2020) To be published
Kronberg ML (1957) Acta Metall 5:507
Kronberg ML (1959) J Nucle Mater 1:85
Kronberg ML (1961) Acta Metall 9:970
Kumar G, Desai A, Schroers J (2010) Adv Mater 1
Kumar KS, Hazzledine PM (2004) Intermetallics 12:763
Laplanche G, Kosta A, Reinhart C, Hunfeld J, Eggeler G, George EP (2017) Acta Mater 128:292
Meng F-Q, Tsuchiya K, Ii S, Yokoyama Y (2012) Appl Phys Let 101:121914
Moriwaki M, Ito K, Inui H, Yamaguchi M (1997) Mater Sci Eng A239–240:63
Nabarro FRN (1967) Theory of crystal dislocations. Clarendon, Oxford
Ogata S, Shimizu F, Li J, Wakeda M, Shibutani Y (2006) Intermetallics 14:1033
Okamoto NL, Kashioka D, Inomoto M, Inui H, Takebayashi H, Yamaguchi S (2013) Scripta Mater 69:307
Okamoto NL, Inomoto M, Adachi H, Takebayashi H, Inui H (2014) Acta Mater 65:229
Okamoto NL, Fujimoto S, Kambara Y, Kawamura M, Chen ZMT, Matsunoshita H, Tanaka K, Inui H, George EP (2016) Sci Rep 6:35863
Otto F, Yang Y, Bei H, George EP (2013) Acta Mater 61:2628
Read-Hill RE, Hirth JP, Rogers HG (eds) (1963) Deformation twinning. Gordon and Breach, New York
Tanaka I, Togo A, this book
Tsuji N, Ogata S, Inui H, Tanaka I, Kishida K, Gao S, Mao W-Q, Bai Y, Zheng R-X, Du J-P (2020) Scripta Mater 181:35
Uchic MD, Dimiduk DM, Florando JN, Nix WD (2004) Science 305:986
Venables JA (1961) Phil Mag 6:379
Weertman J, Weertman JR (1964) Elementary dislocation theory. The McMillan, New York
Yashiro K (2012) JSME 2th computational mechanics conference (in Japanese)
Zhang J, Kishida K, Inui H (2017) Int J Plast 92:45

Chapter 7
TEM Characterization of Lattice Defects Associated with Deformation and Fracture in α-Al₂O₃

Eita Tochigi, Bin Miao, Shun Kondo, Naoya Shibata, and Yuichi Ikuhara

7.1 Introduction

Alumina (α-Al$_2$O$_3$) is structural ceramics used for various applications such as refractory material, thermal coating, and film substrates. The mechanical behavior of alumina has been extensively investigated for decades. Discussion on microstructural plastic deformation mechanisms for single-crystalline alumina (sapphire) was started by Kronberg in 1957 (Kronberg 1957), who proposed dislocation behavior

E. Tochigi (✉) · B. Miao · S. Kondo · N. Shibata · Y. Ikuhara
Institute of Engineering Innovation, The University of Tokyo, 2-11-16 Yayoi, Bunkyo-ku, Tokyo 113-8656, Japan
e-mail: tochigi@sigma.t.u-tokyo.ac.jp

B. Miao
e-mail: miao@sigma.t.u-tokyo.ac.jp

S. Kondo
e-mail: kondou@sigma.t.u-tokyo.ac.jp

N. Shibata
e-mail: shibata@sigma.t.u-tokyo.ac.jp

Y. Ikuhara
e-mail: ikuhara@sigma.t.u-tokyo.ac.jp

E. Tochigi
PRESTO, Japan Science and Technology Agency, 4-1-8, Honcho, Kawaguchi, Saitama 332-0012, Japan

N. Shibata · Y. Ikuhara
Nanostructures Research Laboratory, Japan Fine Ceramics Center, 2-4-1, Mutsuno, Atsuta-ku, Nagoya 456-8587, Aichi, Japan

S. Kondo · Y. Ikuhara
Center for Elements Strategy Initiative for Structural Materials (ESISM), Kyoto University, Yoshidahonmachi, Sakyo-ku, Kyoto 606-8501, Japan

© The Author(s) 2022
I. Tanaka et al. (eds.), *The Plaston Concept*,
https://doi.org/10.1007/978-981-16-7715-1_7

for the basal slip and basal twinning based on crystallographic considerations. Experimental characterizations of dislocations in deformed alumina crystals were actively performed by transmission electron microscopy (TEM) from the 1970s to 1990s (Pletka et al. 1974; Bilde-Sørensen et al. 1976, 1996; Mitchell et al. 1976; Firestone and Heuer 1976; Lagerlöf et al. 1984, 1994), and their structural details, such as Burgers vectors and dissociation reactions, have been revealed. These studies deepened my understanding of the plastic deformation processes of sapphire based on dislocation behavior. Since plastic deformation essentially corresponds to atomic rearrangement due to applied load, the atomistic behavior during plastic deformation has often been discussed, even by Kronberg. However, experimental evidence on atomic structures had not been obtained for years because of the lack of imaging techniques. With the development of TEM techniques, atomic-scale observations of dislocations in alumina were eventually realized since the 2000s, and their core atomic structures have been characterized so far (Nakamura et al. 2002, 2006; Ikuhara et al. 2003; Shibata et al. 2007; Heuer et al. 2010; Tochigi et al. 2010, 2011, 2012, 2015, 2016, 2017, 2018; Miao et al. 2019). Note that some of these studies examined dislocations in low-angle grain boundaries. The dislocation character characterized by Burgers vector and line direction can be controlled through the grain boundary orientation, and a dislocation equivalent to one introduced by plastic deformation can be easily obtained.

In situ TEM indentation experiment is a powerful technique to investigate microstructural evolution upon loading. Modern indentation holders for TEM are driven by a piezo actuator, and the indenter tip can be controlled in the sub-nanometer step (ex. PI-95, Bruker Corp.). To apply a load at a specific local area in a sample, deformation or fracture phenomena can be induced in a controlled manner. In general, in situ TEM indentation is performed in a conventional TEM because of the limitation of equipment compatibility. The experimental information is limited to the nanometer scale. Thus, post-mortem characterization by atomic-resolution TEM is often carried out to obtain detailed information at the atomic level. For alumina, crack propagation (Sasaki et al. 2012; Kondo et al. 2019) and dislocation formation phenomena (Miao et al. 2019) have been investigated by in situ indentation and atomic structure analysis so far.

In this report, we review our recent progress on structural analysis of lattice defects in alumina by TEM observations. In Sect. 7.2, the structures of dislocations with the Burgers vector $b = 1/3 <11\bar{2}0>$, $<1\bar{1}00>$, and $1/3 <\bar{1}101>$ formed in low-angle grain boundaries are discussed based on experimental observations. In Sect. 7.3, the formation of a $1/3 <11\bar{2}0>$ mixed basal dislocation and fracture of Zr-doped $\{1\bar{1}04\}/<11\bar{2}0> \sum 13$ grain boundary are demonstrated by in situ TEM indentation experiments. These phenomena are further investigated at the atomic level based on atomic-resolution scanning TEM (STEM) observations performed after the indentation experiments.

7.2   Atomic Structure Analysis of Dislocations in Low-angle Boundaries

The plastic deformation of alumina is mainly produced by dislocation slip at elevated temperatures. The slip systems of alumina are known to be $(0001)1/3 <11\bar{2}0>$ basal slip (Kronberg 1957; Pletka et al. 1974; Mitchell et al. 1976; Lagerlöf et al. 1984, 1994; Bilde-Sørensen et al. 1996; Nakamura et al. 2002; Shibata et al. 2007; Heuer et al. 2010; Miao et al. 2019), $\{11\bar{2}0\}<1\bar{1}00>$ prism-plane slip (Bilde-Sørensen et al. 1976; Lagerlöf et al. 1984), and $\{1\bar{1}02\}1/3<\bar{1}101>$ ($\{10\bar{1}1\}$ and $\{2\bar{1}\bar{1}3\}$ planes are also possible to be slip planes) pyramidal slip (Firestone and Heuer 1976). Thus, the $1/3<11\bar{2}0>$, $<1\bar{1}00>$, and $1/3<\bar{1}101>$ dislocations play an important role in the plastic deformation of alumina. Conventional TEM studies revealed that these dislocations are basically dissociated into some partial dislocations with stacking faults in between (Bilde-Sørensen et al. 1976; Mitchell et al. 1976; Lagerlöf et al. 1984, 1994). So far, their detailed atomic structures had not been well characterized.

Atomic-resolution TEM is a powerful technique to directly observe the atomic structure of a dislocation. Since this method provides a projected image of three-dimensional structures, the observed dislocation must be straight at end-on orientation. However, such dislocations are rarely found in a deformed crystal. Instead, low-angle grain boundaries can be used. They are divided into tilt and twist boundaries, which consist of an edge dislocation array and a screw dislocation network, respectively. The Burgers vector and line direction of the dislocations depend on the grain boundary orientation. Basically, an edge dislocation in a low-angle tilt grain boundary has the Burgers vector corresponding to the translation vector perpendicular to the boundary plane and line direction parallel to the rotation axis. A screw dislocation in a low-angle twist grain boundary has the Burgers vector and line direction corresponding to the smallest translation vector on the boundary plane. Based on these relationships, dislocation structures can be controlled through grain boundary orientation.

To obtain $1/3<11\bar{2}0>$, $<1\bar{1}00>$, and $1/3<\bar{1}101>$ dislocations, alumina bicrystals with the $\{11\bar{2}0\}/<1\bar{1}00>$, $(0001)/[0001]$, $(1\bar{1}00)/<11\bar{2}0>$, and $(0001)/<1\bar{1}00>$, low-angle grain boundary were fabricated by joining two pieces of single crystals at 1500 °C for 10 h in air. Thin foil samples with a grain boundary for TEM observations were prepared from the bicrystals by cutting, mechanical grinding, and Ar ion milling. The samples were observed by using conventional TEM (JEM-2010HC, 200 kV, JEOL), high-resolution TEM (HRTEM: JEM-4010, 400 kV, JEOL), and scanning TEM (STEM: ARM-200F, 200 kV, JEOL).

7.2.1 $1/3<11\bar{2}0>$ Basal Edge Dislocation

$1/3<11\bar{2}0>$ vector is the translation vector perpendicular to the $\{11\bar{2}0\}$ plane. It is considered that the $\{11\bar{2}0\}$ low-angle tilt grain boundary consists of $1/3<11\bar{2}0>$

edge dislocations. Figure 7.1 shows a bright-field TEM image of the $\{11\bar{2}0\}<1\bar{1}00>$ 2° low-angle tilt grain boundary. Dark contrast pairs periodically array in line with a separation distance of ~13 nm, suggesting that the $1/3 <11\bar{2}0>$ dislocation dissociates into two partial dislocations with a stacking fault in between. The tilt angle of the grain boundary 2θ and the dislocation configuration are related to Frank's formula, $2\theta = b/d$ (Frank 1951), where b is the magnitude of the Burgers vector of the perfect dislocation (0.476 nm) and d is the distance between the perfect dislocations. The tilt angle is estimated to be 2.1°, which agrees with the designed value. Previous TEM studies and crystallographic considerations revealed that the dissociation reaction of $1/3 <11\bar{2}0>$ dislocation follows the equation below (Mitchell et al. 1976; Lagerlöf et al. 1984, 1994; Ikuhara et al. 2003; Nakamura et al. 2006; Tochigi et al. 2008):

$$1/3 < 11\bar{2}0 > \rightarrow 1/3 < 10\bar{1}0 > +1/3 < 01\bar{1}0 > . \qquad (7.1)$$

Figure 7.2 shows an HRTEM image of the $1/3 <11\bar{2}0>$ dislocation viewed along $<1\bar{1}00>$ direction. The two extra-half planes are located along [0001] direction with a separation distance of about 3 nm. This indicates that the dislocation consists of two partial dislocations with a stacking fault on $\{11\bar{2}0\}$ plane, although the stacking

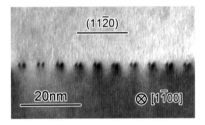

Fig. 7.1 TEM image of the $\{11\bar{2}0\}/<1\bar{1}00>$ 2° tilt grain boundary of alumina. Dislocation pairs periodically array along the grain boundary (Nakamura et al. 2006; Tochigi et al. 2008)

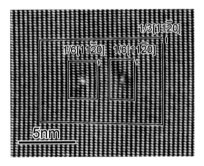

Fig. 7.2 HRTEM image of a dislocation pair viewed along [1$\bar{1}$00] direction. The perfect dislocation dissociates into two partial dislocations with a stacking fault on (11$\bar{2}$0) plane. The Burgers circuit indicates that the perfect dislocation has an edge component of 1/3[11$\bar{2}$0] and the partial dislocations have 1/6[11$\bar{2}$0] (Nakamura et al. 2006)

fault appears like the bulk in the present [$1\bar{1}00$] projection. Since the two partial dislocations are on different slip planes, this dissociation is found to occur by climb mechanism with atomic diffusion. Theoretical studies suggested that the formation energy of stacking faults (stacking fault energy) on the (0001) plane is at least one order higher than that on the {$11\bar{2}0$} plane (Y. 2008; Kenway 1993; Marinopoulos and Elsässer 2001). This would be the reason why the 1/3 <$11\bar{2}0$> edge dislocation does not dissociate on the (0001) plane by glide mechanism. The Burgers circuit shows the edge component of the partial dislocations is 1/6 <$11\bar{2}0$>. This is consistent with the partial dislocations that have the Burgers vector of 1/3 <$10\bar{1}0$> and 1/3 <$01\bar{1}0$>, the edge components of which are 1/6 <$11\bar{2}0$>. The formation energy of the stacking fault (stacking fault energy) formed between two partial dislocations can be estimated from the separation distance based on the linear elastic theory. For a low-angle grain boundary, contributions from all the dislocations depending on their configurations need to be taken into account. In the present case, the stacking fault energy (γ) is calculated by the following equation (Ikuhara et al. 2003; Jhon et al. 2005):

$$\gamma = \frac{\mu b_p^2 (2 + \nu)}{8\pi (1 + \nu)} \cdot \frac{1}{d} \sum_{n=0}^{\infty} \left(\frac{1}{n + \alpha} - \frac{1}{n + 1 - \alpha} \right), \tag{7.2}$$

where μ is the shear modulus (~150 GPa (Nakamura et al. 2009)), b_p is the magnitude of the Burgers vector of the partial dislocations (0.275 nm), ν is Poisson's ratio (~0.24 (Chung and Simmons 1968)), and α is d_1/d (d_1: the spacing between the partial dislocations). Substituting the averaged values of $d = 13.2$ nm and $d_1 = 4.6$ nm into Eq. 7.2, the formation energy was estimated to be 0.32 Jm^{-2}, which agrees with the values estimated by HRTEM observations of a basal dislocation in a deformed crystal (0.28 Jm^{-2} (Nakamura et al. 2002)) and by first-principles calculations (0.35 Jm^{-2} (Marinopoulos and Elsässer 2001)). It can be said that analysis of low-angle boundaries is useful to obtain structural information of dislocation.

7.2.2   1/3 <$11\bar{2}0$> Basal Screw Dislocation

Three equivalent translation vectors of 1/3 <$11\bar{2}0$> exist on the (0001) basal plane, and thus a (0001) low-angle twist grain boundary should consist of a network of 1/3 <$11\bar{2}0$> screw dislocations. Figure 7.3 shows a plan-view bright-field TEM image of a (0001)/[0001] low-angle twist grain boundary (Tochigi et al. 2012). A hexagonal dislocation network is observed. Since the dislocation line directions are parallel to either of [$11\bar{2}0$], [$1\bar{2}10$], and [$\bar{2}110$] directions, the dislocation network consists of the 1/3 <$11\bar{2}0$> screw dislocations. The periodicity of equivalent dislocations (d) is about 60 nm. The twist angle (φ) can be calculated to be 0.45°. To characterize the core atomic structure of the 1/3 <$11\bar{2}0$> screw dislocation, a <$11\bar{2}0$> cross-sectional sample was prepared and observed by HRTEM. Figure 7.4 shows the core atomic structure of the 1/3 <$11\bar{2}0$> screw dislocation. A little contrast

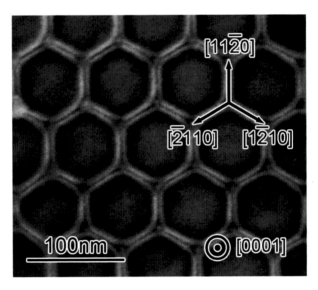

Fig. 7.3 Plan-view TEM image of the (0001)/[0001] twist grain boundary of alumina. A hexagonal dislocation network is formed. The line direction of dislocations is either of [11$\bar{2}$0], [1$\bar{2}$10], and [$\bar{2}$110], indicating that they are 1/3 <11$\bar{2}$0> screw dislocations (Tochigi et al. 2012)

Fig. 7.4 HRTEM image of 1/3 <11$\bar{2}$0> screw dislocation viewed end on. The lattice disorder at the center corresponds to the dislocation core. The circuit without disconnection shows that it is the perfect screw dislocation (Tochigi et al. 2012)

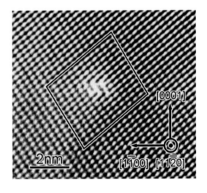

disorder appears at the center without any lattice discontinuities as shown by the closed circuit, indicating the presence of the screw dislocation. This screw dislocation is in the perfect type in contrast to the basal edge dislocation consisting of the partial dislocation pair. If the screw dislocation dissociates by climb mechanisms following Eq. 7.1, the accumulation of extra atoms or vacancies between partial dislocations is structurally required. This process may need high energy and is difficult to occur. A more detailed discussion on the dissociation mechanisms of basal dislocations will be given in Sect. 7.3.1.

7.2.3 <1$\bar{1}$00> Edge Dislocation

Figure 7.5 shows a bright-field TEM image of a {1$\bar{1}$00}/<11$\bar{2}$0> low-angle 2° tilt grain boundary (Tochigi et al. 2010, 2016). Dark contrast triplets periodically array in line. Since the translation vector normal to the {1$\bar{1}$00} plane is <1$\bar{1}$00>, the dark contrast triplet is originated from the 1/3 <1$\bar{1}$00> partial dislocation triplets formed by the following dissociation reaction:

$$< 1\bar{1}00 > \rightarrow 1/3 < 1\bar{1}00 > +1/3 < 1\bar{1}00 > +1/3 < 1\bar{1}00 > . \qquad (7.3)$$

Figure 7.6 shows an HRTEM image of the 1/3 <1$\bar{1}$00> partial dislocation triplet. The partial dislocations are located along the [0001] direction with separation distances of approximately 10.1 and 10.9 nm from the left-hand side. Two stacking faults are formed on the {1$\bar{1}$00} plane, and they are structurally inequivalent (Lagerlöf et al. 1984). As a result of detailed analysis, it was found that the left stacking fault has the stacking sequence of …ABCCABC… referred to as I$_2$ type, and the right one has …ABCBCABC… referred to as V type (Tochigi et al. 2010). In the present case, their stacking fault energies can be calculated by the following equations:

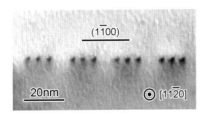

Fig. 7.5 TEM image of the {1$\bar{1}$00}/<11$\bar{2}$0> 2° tilt grain boundary of alumina. The grain boundary consists of dislocation triplets (Tochigi et al. 2010)

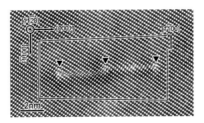

Fig. 7.6 HRTEM image of a dislocation triplet view along [11$\bar{2}$0]. The circuit shows that the perfect dislocation has the Burgers vector of[1$\bar{1}$00]. The stacking faults are formed on {1$\bar{1}$00} plane. A detailed analysis found that the left stacking fault has the staking sequence of …ABCCABC… (I$_2$ type) and the right one has …ABCBCABC… (V type) (Tochigi et al. 2010)

$$\gamma_{I2} = \frac{\mu b_p^2}{2\pi(1-\nu)} \cdot \frac{1}{d} \sum_{n=0}^{\infty} \left(\frac{1}{n+\alpha_1} + \frac{1}{n+\alpha_1+\alpha_2} - \frac{1}{n+1-\alpha_1} - \frac{1}{n+1-\alpha_1-\alpha_2} \right), \quad (7.4)$$

$$\gamma_V = \frac{\mu b_p^2}{2\pi(1-\nu)} \cdot \frac{1}{d} \sum_{n=0}^{\infty} \left(\frac{1}{n+\alpha_2} + \frac{1}{n+\alpha_1+\alpha_2} - \frac{1}{n+1-\alpha_2} - \frac{1}{n+1-\alpha_1-\alpha_2} \right), \quad (7.5)$$

where α_1 is d_1/d (d_1: the width of stacking fault I_2) and α_2 is d_2/d (d_2: the width of stacking fault V). Substituting the experimental averaged values of $d = 22$ nm, $d_1 = 4.9$ nm, and $d_2 = 5.7$ nm, into Eqs. (7.4) and (7.5), the stacking fault energies $\gamma_{I2} = 0.44$ Jm^{-2} and $\gamma_V = 0.35$ Jm^{-2}, which agree with the values estimated by theoretical studies using first-principles calculations, 0.46 and 0.41 Jm^{-2}. Note that the $<1\bar{1}00>$ edge dislocation associated with $\{11\bar{2}0\}<1\bar{1}00>$ slip has the line direction along the [0001] direction and is not structurally equivalent to the dislocation characterized here. The slip dislocation is also known to dissociate into three partial dislocations with $\{1\bar{1}00\}$ stacking faults following Eq. (7.3) (Bilde-Sørensen et al. 1976; Lagerlöf et al. 1984). These stacking faults should be I_2 and V types, although no direct evidence has ever been obtained. This is because the other possible $\{1\bar{1}00\}$ stacking fault ...ABCBABC... referred to as I_1 type is calculated to have the formation energy of 0.62–0.63 Jm^{-2}, which is about 1.4–1.8 times higher than that of I_2 and V types (Tochigi et al. 2010; Marinopoulos and Elsässer 2001).

7.2.4   1/3 $<\bar{1}101>$ *Mixed Dislocation*

Figure 7.7 shows a TEM image of the $(0001)/<1\bar{1}00>$ 2° tilt grain boundary (Tochigi et al. 2015). Bright discrete contrasts are observed along the grain boundary, suggesting that dislocations array periodically. An HRTEM image of the grain boundary is shown in Fig. 7.8. Dislocation contrasts are seen, and each dislocation is elongated on the (0001) plane with a separation distance of about 4 nm, indicating that the dislocation dissociates into two partial dislocations with a (0001) stacking fault. The Burgers circuits drawn around the dislocations indicate that the dislocations have the edge component of either 1/3[0001], 1/6[$\bar{1}\bar{1}22$], or 1/6[11$\bar{2}2$]. Since these components do not coincide with a translation vector, the dislocations should be the mixed type having a screw component along the [1$\bar{1}$00] direction. The

Fig. 7.7 TEM image of the $(0001)/<1\bar{1}00>$ 2° tilt grain boundary of alumina. Bright contrasts are periodically seen along the boundary (Tochigi et al. 2015)

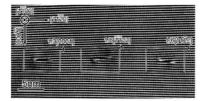

Fig. 7.8 HRTEM image of the grain boundary viewed edge on. Three dislocations having an edge component of 1/3[0001], 1/6[$\bar{1}$122], and 1/6[11$\bar{2}$2] are observed. Each dislocation dissociates into two partial dislocations with stacking fault on (0001) plane (Tochigi et al. 2015)

screw components can be identified based on crystallography. For the case of the edge component of 1/6[$\bar{1}$122], considering the formula 1/6[$\bar{1}$122] + n[1$\bar{1}$00], it equals to the smallest translation vector of 1/3[0$\bar{1}$11] for $n = 1/6$. Thus, the screw component and the total Burgers vector are identified to be 1/6[1$\bar{1}$00] and 1/3[0$\bar{1}$11], respectively. Similarly, for the cases of 1/6[11$\bar{2}$2], and 1/3[0001], the total Burgers vectors were found to be 1/3[10$\bar{1}$1] and 1/3[$\bar{1}$101], respectively. The sum of 1/3[0$\bar{1}$11], 1/3[10$\bar{1}$1], and 1/3[$\bar{1}$101] is [0001], which is the translation vector perpendicular to the (0001) grain boundary plane. In the present boundary, it was found that three equivalent mixed dislocations accommodate the grain boundary misorientation.

Figure 7.9 shows an annular bright-field (ABF) STEM image of the 1/3[0$\bar{1}$11] dislocation. In this image, dark contrasts correspond to atomic columns as indicated by the atomic model overlaid. The edge components of the partial dislocations are found to be 1/18[$\bar{1}$123] and 1/18[$\bar{2}\bar{2}$43], and the stacking fault formed on the (0001) plane has the stacking sequence of …2B 1 2C 3 1A 2 1B 3…. The Burgers vector of the partial dislocations can be estimated through the fault vector of the stacking fault. A detailed analysis using first-principles calculations revealed the fault vector

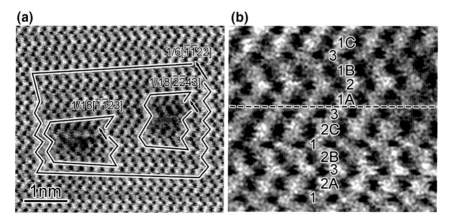

Fig. 7.9 **a** ABF-STEM image of the dislocation with the edge component of 1/6[$\bar{1}$122]. The edge component of two partial dislocations are 1/18[$\bar{2}$4$\bar{2}$3] and 1/18[$\bar{2}\bar{2}$43]. **b** Enlarged image of (0001) stacking fault. Its stacking sequence is …2B 1 2C 3 1A 2 1B 3… (Tochigi et al. 2015)

of the stacking fault, and the Burgers vectors of the partial dislocations were found to be $1/18[2\bar{4}23]$ and $1/18[\bar{2}\bar{2}43]$ (Tochigi et al. 2015). The dissociation reaction of $1/3 <0\bar{1}11>$ dislocation is described by the following equation:

$$1/3 < 0\bar{1}11 > \rightarrow 1/18 < 2\bar{4}23 > +1/18 < \bar{2}\bar{2}43 > . \qquad (7.6)$$

It was confirmed that the dissociation of the other two cases is also given by Eq. (7.6) (Tochigi et al. 2015). The stacking fault energy was estimated to be experimentally 0.58 Jm^{-2} and theoretically 0.72 Jm^{-2}, and they agree with each other. The $1/3 <0\bar{1}11>$ dislocation was found to dissociate on the (0001) plane, not corresponding to the slip plane. Therefore, the $1/3 <0\bar{1}11>$ dislocations may become immobile by the dissociation reaction.

7.3 Analysis of Dislocation Formation and Grain Boundary Fracture by in Situ TEM Nanoindentation and Atomic-Resolution STEM

To directly observe the dynamic behavior of lattice defects upon loading, in situ TEM mechanical experiment is an efficient method. In this section, the formation process of a basal dislocation (Miao et al. 2019) and the propagation process of a crack along a large angle grain boundary of alumina (Kondo et al. 2019) are demonstrated by an in situ TEM nanoindentation experiment. Furthermore, the nucleated defects were statistically characterized by STEM after the nanoindentation experiment to develop an atomistic understanding of their formation processes.

7.3.1 *Introduction of a Basal Mixed Dislocation and Its Core Structure*

The $(0001)1/3 <1\bar{2}10>$ basal slip becomes to be the easiest slip system at elevated temperature $>\sim 1000$ °C (Lagerlöf et al. 1994). In addition, the core structures of the $1/3 <1\bar{2}10>$ basal dislocation in different orientation have been investigated for approximately 20 years. So far, atomic-resolution TEM/STEM studies characterized the dislocation core structures in edge (Nakamura et al. 2002, 2006; Shibata et al. 2007; Heuer et al. 2010), 30° (Heuer et al. 2010), and screw (Tochigi et al. 2012) types. However, the core structure of 60° mixed dislocation has not been observed yet. Here, we demonstrate the formation process and core structure of the 60° mixed dislocation (Miao et al. 2019).

An alumina single crystal plate was mounted on a half-moon-shaped mesh and thinned by mechanical grinding. The sample was further milled by Ar ion milling until its free edge obtains electron transparency. The face of the sample was parallel

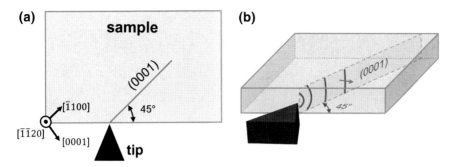

Fig. 7.10 Schematic illustrations showing the crystallographic orientation of the sample and the indentation direction. **a** The indentation direction was set at an angle of 45° from (0001) plane. **b** 3D illustration. The red arrow indicates the Burgers vector of dislocation expected to be induced by indentation, which is $1/3[1\bar{2}10]$ (Miao et al. 2019)

to the $(11\bar{2}0)$ plane and the free edge was set at an angle of 45° from the (0001) plane, as illustrated in Fig. 7.10. The indentation direction was set perpendicular to the free edge. Assuming a unidirectional force, the $(0001)1/<1\bar{2}10>$ basal slip has the largest Schmid factor of 0.5. In situ TEM nanoindentation experiment was performed using a nanoindentation TEM sample holder (Nanofactory Instruments AB, Sweden) in a conventional TEM (JEM-2100HC, 200 kV, JEOL). After the nanoindentation experiment, the sample was further thinned by Ar ion milling at liquid nitrogen temperature to get the deformed area thin. The core atomic structure of introduced dislocations was observed by atomic-resolution STEM (ARM-200F, 200 kV, JEOL).

Figure 7.11 shows TEM images taken before and after indentation. The alumina sample is seen at the upper part of the image and the indenter tip is below. Before indentation, no obvious defects are included in the sample (Fig. 7.11a). After indentation, bend contours are seen over the sample and the sample edge is chipped, where a strong indentation force was expected to be applied (Fig. 7.11b). The straight dark contrast is generated from the chipped area and is parallel to (0001), suggesting that basal slip was activated and some basal dislocations were formed there. To characterize the basal dislocations in detail, the sample was mounted on a double tilt sample holder and the indented area was observed with tilt-controlled conditions. Figure 7.11c shows a bright-field TEM image taken along the $[11\bar{2}0]$ zone axis. Two dotted contrast features are seen at the end of the straight contrast. These features were observed along the $[22\bar{4}1]$ direction by tilting the sample by 24.5 degrees around the $[1\bar{1}00]$ axis (Fig. 7.11d). They appear as curved lines and are found to be basal dislocations.

Figure 7.12a shows an ABF-STEM image of one of the dislocations induced by the nanoindentation experiment. The dislocation was observed along the $[11\bar{2}0]$ direction and found to dissociate into two partial dislocations with a separation distance of approximately 2.9 nm. A stacking fault is formed on the $\{1\bar{1}00\}$ plane, and this dissociation also occurred by climb. The largest Burgers circuit shows that

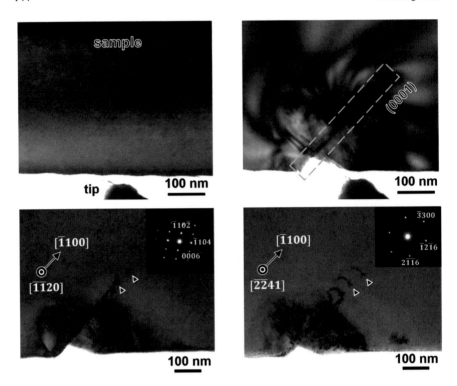

Fig. 7.11 TEM images of the sample before and after the indentation experiment. **a** Before indentation. The sample exhibits a uniform contrast and seems to have no obvious defects. **b** After indentation. The sample edge is chipped due to the insertion of the indenter tip. A straight contrast parallel to (0001) plane from the top of the chipped region, indicating that (0001) slip system was activated. **c** Bright-field image of the indented sample viewing along [11$\bar{2}$0] direction. Two dotted contrasts are seen along (0001) plane. **d** [$\bar{2}\bar{2}$41] view of the sample, where it was rotated by 24.5 degrees around [$\bar{1}$100] axis. The dotted contrasts in (**c**) become curved contrasts, indicating they correspond to basal dislocations (Miao et al. 2019)

the perfect dislocation has an edge component of 1/6[1$\bar{1}$00]. Since this component does not correspond to a translation vector, the perfect dislocation should have a screw component along the [11$\bar{2}$0] direction. From crystallographic considerations, the screw component can be determined in order to get the smallest translation vector, which is 1/6 <11$\bar{2}$0>. Therefore, the perfect dislocation is identified to be 1/3 <1$\bar{2}$10> 60° mixed dislocation. In Fig. 7.12a, the small Burgers circuits indicate that the edge components of the upper and lower partial dislocations are 1/3[1$\bar{1}$00] and 1/6[1$\bar{1}$00], respectively. Since the dissociation reaction follows Eq. (7.1), it is considered that the upper partial dislocation is 1/3[1$\bar{1}$00] edge dislocation and the lower dislocation is 1/3[01$\bar{1}$0] 30° mixed dislocation. The crystallographic orientation of this dissociated dislocation on the (0001) plane is illustrated in Fig. 7.12b. The vectors of 1/3[1$\bar{2}$10], 1/6[1$\bar{1}$00], and 1/6[01$\bar{1}$0] are drawn in the green, blue, and red arrows. The [11$\bar{2}$0] projection of these vectors coincides with the edge component

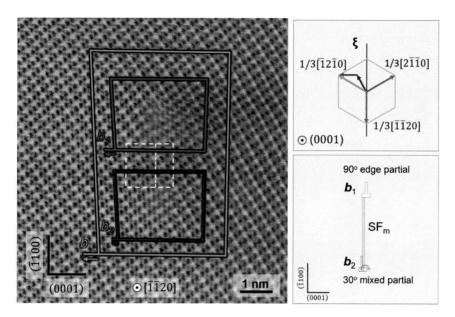

Fig. 7.12 **a** ABF-STEM image of a basal dislocation viewing along [$\bar{1}\bar{1}20$] direction, which was introduced by the indentation experiment. The dislocation dissociates into two partial dislocations with {$1\bar{1}00$} stacking fault by climb. b, b_1, and b_2 correspond to the vector of 1/3 [$\bar{1}2\bar{1}0$], 1/3 [$\bar{1}100$], and 1/3[$01\bar{1}0$], and their edge components 1/2 [$\bar{1}100$], 1/3 [$\bar{1}100$], and 1/6 [$\bar{1}100$] are seen in the image. **b** Schematic illustration showing (0001) projection of the Burgers vectors. Green, blue, and red arrows correspond to b, b_1, and b_2, respectively. The viewing or dislocation line direction is represented by ξ. **c** Schematic illustration showing the dissociated structure of the 1/3 [$\bar{1}2\bar{1}0$] 60° mixed dislocation, which consists of 90° edge and 30° mixed partial dislocations (Miao et al. 2019)

of the dislocations shown in Fig. 7.12a. The dissociated structure observed along [$11\bar{2}0$] is schematically illustrated in Fig. 7.12c.

Figure 7.13 shows magnified images of the bulk and the {$1\bar{1}00$} stacking fault formed between the partial dislocations. The position of the stacking fault is indicated by the dashed line, and structural features are indicated by the circles. In comparison with the experimental image and the structure model, the stacking sequence is identified to be …ABCBCABC… referred to as V type as mentioned in Sect. 7.2.3 (Lagerlöf et al. 1984; Tochigi et al. 2010, 2016; Marinopoulos and Elsässer 2001). The stacking fault energy can be estimated by the following equation:

$$\gamma = \frac{\mu b_p^2 (2 - \nu)}{8 \pi d_1 (1 - \nu)} \left(1 - \frac{2\nu - cos2\beta}{2 - \nu} \right), \qquad (7.7)$$

where β means the angle between the total Burgers vector and dislocation line direction. Substituting the separation distance $d_1 = 2.9$ nm and $\beta = 60°$, the stacking fault energy is 0.41 Jm^{-2}, which agrees well with the experimental and theoretical values of 0.35 Jm^{-2} and 0.41 Jm^{-2} estimated in Sect. 7.2.3.

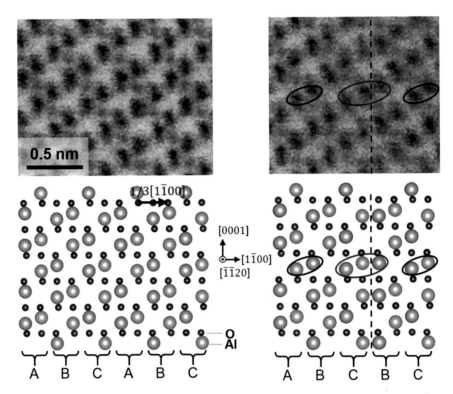

Fig. 7.13 Enlarged ABF-STEM image and atomic structure model of **a** bulk region and **b** {1$\bar{1}$00} stacking fault formed between the partial dislocations. The position of the stacking fault is indicated by dashed lines. The stacking sequence is …ABCBC… corresponding to V-type. The red ellipse shows structural features of the stacking fault (Miao et al. 2019)

The present experiment revealed the core structure of 60° mixed basal dislocation, and thus the core structures of the edge, 30°, 60°, and screw basal dislocations have been experimentally identified, which are summarized in Fig. 7.14. The edge, 60° mixed and 30° mixed basal dislocations dissociate into two partial dislocations following Eq. (7.1). The dissociation of the edge and 60° mixed dislocations occur by climb, while that of the 30° mix dislocation occurs by cross slip because one of the partial dislocations is screw type. Their stacking fault plane is perpendicular to the basal plane, which is the {11$\bar{2}$0} plane for the edge and 30° dislocations and the {1$\bar{1}$00} plane for the 60° mixed dislocation. These dissociation reactions occur by climb rather than glide. A possible reason is that the stacking faults formed on the (0001) plane are predicted to have one order higher energy than the {11$\bar{2}$0} and {1$\bar{1}$00} stacking faults (Y. 2008; Kenway 1993; Marinopoulos and Elsässer 2001). The screw basal dislocation does not dissociate and has the perfect type of core structure. Considering geometric relationship and stability of stacking fault, the screw dislocation is possible to dissociate with the {1$\bar{1}$00} stacking fault in V type as illustrated in Fig. 7.13e. However, to form this dissociated structure, accumulation

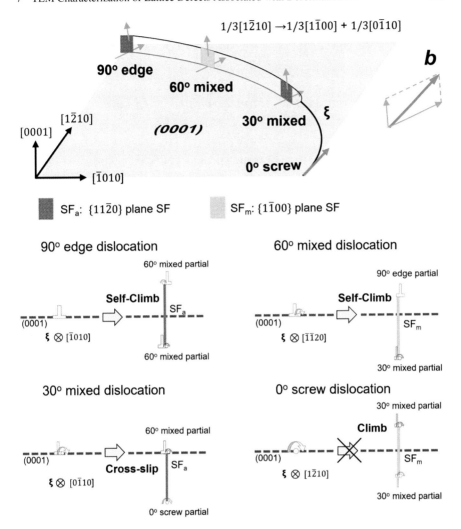

Fig. 7.14 Schematic illustrations showing dissociated structures of a basal dislocation in alumina. **a** A curved $1/3[1\bar{2}10]$ dislocation containing the 90°, 60°, 30°, 0° parts. The former three parts are dissociated into $1/3[1\bar{1}00]$ and $1/3[01\bar{1}0]$ partial dislocations. **b** 90° dislocation. Its dissociation occurred by climb with the $\{1\bar{2}10\}$ stacking fault. The two partial dislocations have 60° orientation. **c** 60° dislocation. Its dissociation occurred by climb with the $\{1\bar{1}00\}$ stacking fault. The two partial dislocations have 90° and 30° orientation. **d** 30° dislocation. Its dissociation occurred by cross-slip with the $\{1\bar{2}10\}$ stacking fault. The two partial dislocations have 60° and 0° orientation. **e** 0° dislocation. From geometrical considerations, this dislocation can dissociate into 30° partial dislocations, but it does not occur experimentally (Miao et al. 2019)

of vacancies between the partial dislocations is necessary, which could be difficult to realize. This would be the reason why the screw dislocation does not dissociate.

7.3.2 Crack Propagation Along Zr-Doped $\sum 13$ Grain Boundary

Polycrystalline materials are often fractured along grain boundaries. This suggests that grain boundaries can preferentially be a crack propagation path, and the atomic bonding of the grain boundaries is weaker than that of the bulk. However, little is known about the atomistic mechanisms of grain boundary fracture phenomena because of the lack of experimental examinations. In this section, we demonstrate the fracture experiment of a well-oriented grain boundary in TEM (Kondo et al. 2019). The Zr-doped $\{1\bar{1}04\}/<11\bar{2}0> \sum 13$ grain boundary of alumina was prepared by the bicrystal method. Crack propagation along the grain boundary was induced and directly observed by the in situ TEM nanoindentation technique. The fracture surfaces were further characterized by atomic-resolution STEM and first-principles calculations. The mechanisms for the grain boundary fracture will be discussed in terms of the atomic bonding of the grain boundary.

Two pieces of alumina single crystal were precisely cut so as to have the orientation relationship of $\{1\bar{1}04\}/<11\bar{2}0> \sum 13$. The bonding surface of one of them was coated by Zr with about 5 nm thickness using an Ar ion spattering machine. The two crystals were joined to be a bicrystal at 1500 °C for 30 h in air. A small plate including the grain boundary was cut from the bicrystal block and mounted on a half-moon-shaped mesh. The sample was thinned by mechanical grinding and Ar ion milling. Nanoindentation experiments were performed using a double-tilt nanoindentation TEM holder (Nanofactory) and a conventional TEM (JEM-2010HC, 200 kV, JEOL). The indenter tip was inserted along the grain boundary to preferentially induce grain boundary fracture. The indented sample was further observed by atomic-resolution STEM (ARM-200F, 200 kV, JEOL) to characterize the atomic structure of the fractured region. The fracture path was further investigated by first-principles calculations. Firstly, a supercell with the Zr-doped Σ13 grain boundary containing 196 atoms was constructed based on experimental observations. This model was optimized using the VAPS program (Gieske and Barsch 1968) under the generalized gradient approximations (Kresse and Furthmüller 1996). A $3 \times 2 \times 1$ k-point mesh and the energy cutoff of 500 eV were used. Secondly, atomic models of three kinds of fracture surfaces were constructed in order to remove half of the atoms from the optimized supercell. The total energies of these supercells were calculated without structural optimization to estimate the cleavage energies. Thirdly, the structural optimization was carried out for these supercells, and the optimized structures were compared with an experimental image.

Figure 7.15 shows sequential TEM images captured from a movie taken during the nanoindentation experiment. Figure 7.15a shows an image before indentation.

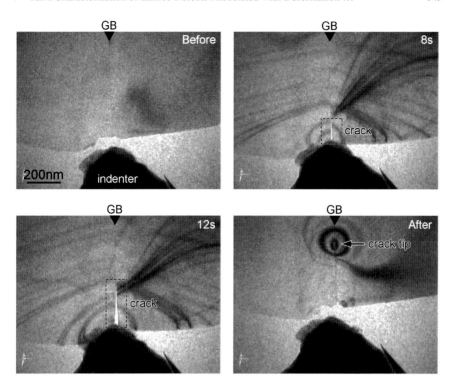

Fig. 7.15 Sequential images captured from a movie taken during in situ TEM nanoindentation experiment. **a** Before indentation. The grain boundary is located vertically in the middle of the sample. **b** After 8 s. The indentation started. A crack initiated along the grain boundary from the sample edge. **c** After 12 s. The crack propagated. **d** After the indenter retracted. Characteristic rounded contrasts are seen in the sample, corresponding to bend contours. The crack tip is expected to be located at their center (Kondo et al. 2019)

The sample and indenter tip are observed. The grain boundary is located vertically in the middle. In 8 s (Fig. 7.15b), a crack is found to nucleate from the top of the indenter tip as indicated by the dashed box in the figure. In 12 s, the crack propagates along the grain boundary (Fig. 7.15c). Figure 7.15d shows a situation after the indentation experiment. The rounded contrasts are seen as pointed by the arrow. They should be bend contours, and thus the crack tip is expected to have propagated around the center of the rounded contrasts.

The indented sample was directly transferred to atomic-resolution STEM. Figure 7.16 shows a high-angle annular dark-field (HAADF) STEM image of the vicinity of crack tip (Fig. 7.16a) and one of the fracture surfaces (Fig. 7.16b). In the HAADF images, strong and weak spots correspond to Zr and Al columns, respectively. In the upper part of Fig. 7.16a corresponding to the unfractured area, three layers of Zr are formed along the grain boundary, namely that Zr forms the segregation structure of three atomic layers in the $\{1\bar{1}04\}/<11\bar{2}0>\sum 13$ grain boundary of alumina. In the lower part, the Zr segregation layers appear a little fuzzy and consist

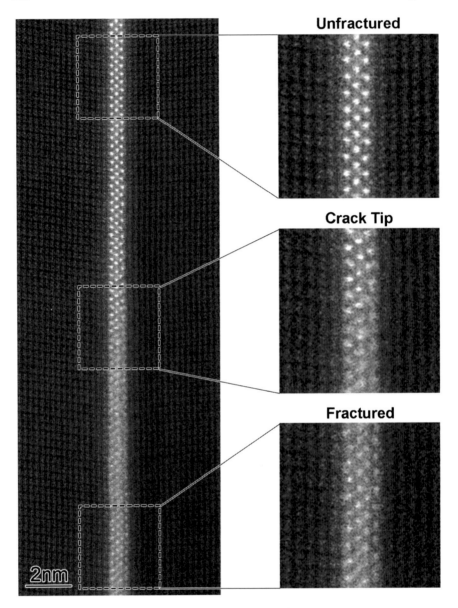

Fig. 7.16 HAADF-STEM image at the crack tip region. The bright contrasts correspond to Zr columns. In the unfractured region at the upper side, a Zr segragation phase in three layers is formed. The fractured region at the lower side appears to consist of four Zr layers, suggesting that the crack propagated within the Zr segregation phase and was divided into the left and right crystals. At the center of the image, contrasts of the Zr layer gradually change. The crack tip is expected to be located there (Kondo et al. 2019)

of four atomic layers, suggesting that the crack propagated within the Zr-segregated phase and the two crystals are slightly apart from each other with Zr layers on their surface. The contrast features of the Zr-segregated phase gradually change across the image center, and thus the crack tip is expected to be located around the image center.

Figure 7.17 shows a HAADF-STEM image of the fracture surface of the left crystal. The two layers of the Zr-segregated phase are observed. As seen in the intensity graph in the below panel, the top layer has a lower intensity than the second layer, although the two layers have a similar intensity in the unfractured area. Since the image intensity of an atomic column in HAADF-STEM depends on its atomic density, the number of atoms in the top layer should have decreased. That is likely to be half, because the left and the right crystals are symmetric. Therefore, it is considered that the crack propagated through the center layer of the Zr-segregated phase so as to equally distribute the center layer into the two symmetric crystals.

To investigate fracture mechanisms of the Zr-doped $\Sigma13$ grain boundary, three kinds of fracture paths for the grain boundary were examined by first-principles calculations. Figure 7.18a shows an optimized structure model of the Zr-doped $\Sigma13$ grain boundary viewing along the orthogonal directions of $[11\bar{2}0]$ and $[02\bar{2}1]$. Note that the structure model agreed well with the experimental image of the grain boundary. The dashed lines indicate three possible fracture paths, Zr-Al (straight), Zr-Zr (straight), and Zr-Zr (zigzag). The oxygen atoms are distributed so as not to produce a high-energy polar surface. The cleavage energies corresponding to these fracture paths were evaluated to be 4.66 Jm^{-2}, 3.61 Jm^{-2}, and 2.66 Jm^{-2}, respectively, indicating that the Zr-Zr (zigzag) path is energetically favorable. A comparison of an experimental image and optimized fracture surfaces formed by the Zr-Zr (zigzag) path and by Zr-Zr (straight) path is shown in Fig. 7.18b. The atomic positions of the Zr-Zr (zigzag) surface agree well with the experimental image, whereas those of the Zr-Zr (straight) surface do not. Therefore, the Zr-Zr (zigzag) path is energetically and structurally reasonable. It is considered that the cleavage energy is related to the atomic bonding of the grain boundary cut by a crack. The fracture due to Zr-Zr (straight) path produces 8 six-fold coordinated Zr on the two fracture surfaces, whereas Zr-Zr (zigzag) path produces 4 six-fold coordinated Zr and 8 seven-fold Zr (Kondo et al. 2019). A theoretical study suggested that zirconium oxides tend to have seven-fold or eight-fold Zr, and six-fold Zr requires relatively higher energy (Perdew et al. 1992). This would be the reason why the Zr-Zr (zigzag) path is selected for the grain boundary fracture.

7.4 Summary

In this report, we reviewed the atomic structure analysis of dislocations and grain boundary fracture surfaces of alumina by TEM/STEM. Using bicrystals with a low-angle grain boundary, $1/3 <11\bar{2}0>$ edge, $<1\bar{1}00>$ edge and $1/3 <\bar{1}101>$ mixed, and $1/3 <11\bar{2}0>$ screw dislocations were fabricated. The former three dislocations

Fig. 7.17 **a** HAADF-STEM image of the fracture surface of the left crystal. Two Zr layers remain on the alumina crystal. **b** The intensity profile of the HAADF-STEM image in (**a**), where signals were integrated along the vertical direction. The intensity of the first layer is lower than the second layer, indicating that Zr atoms in the first layer were divided into two crystals (Kondo et al. 2019)

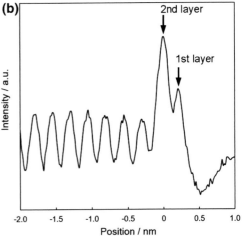

Fig. 7.18 **a** Atomic structure model of Zr-segregated $\{1\bar{1}04\}/<11\bar{2}0>\sum 13$ grain boundary of alumina. The upper and lower panels correspond to $[2\bar{1}\bar{1}0]$ and $[02\bar{2}1]$ projection, respectively. Fracture paths considered in the present study, Zr-Al, Zr-Zr (straight) and Zr-Zr (zigzag), are indicated by dashed lines. **b** Comparison of averaged experimental image and calculated fracture surfaces of Zr-Zr (straight) and Zr-Zr (zigzag). The Zr-Zr zigzag fracture surface agrees well with the experimental image (Kondo et al. 2019)

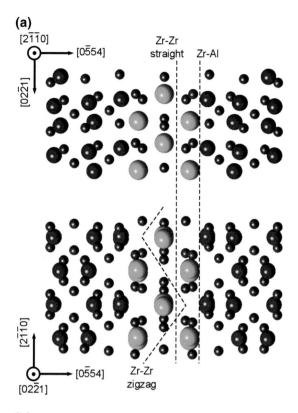

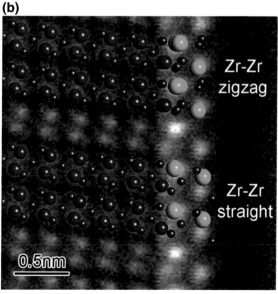

were found to be dissociated into partial dislocations with a stacking fault, whereas the last one has the perfect core. TEM/STEM analysis and theoretical calculations estimated the formation energy of $\{11\bar{2}0\}$, $\{1\bar{1}00\}$, and (0001) stacking faults to be approximately 0.3 Jm^{-2}, 0.4–0.5 Jm^{-2}, and 0.6–0.7 Jm^{-2}, respectively. Except for the $1/3 <11\bar{2}0>$ screw dislocation. In situ TEM nanoindentation technique was applied to induce basal slip in alumina. The 60° mixed $1/3 <11\bar{2}0>$ dislocation on the (0001) plane was introduced, and its core structure was characterized by STEM. It was found that the dislocation consists of two partial dislocations with $\{1\bar{1}00\}$ stacking fault. By this result, the core structures of representative segments (0°, 30°, 60°, and 90°) of the $1/3 <11\bar{2}0>$ basal dislocation were fully revealed. In addition, fracture of the Zr-doped $\{1\bar{1}04\}/<11\bar{2}0> \sum 13$ grain boundary was induced by in situ TEM nanoindentation, and the fracture surfaces were characterized by STEM. It was found that the crack propagated along a zigzag path in order to equally divide the Zr-segregated phase into three atomic layers. Experimental image and theoretical calculations revealed that the fracture along the zigzag path forms stable Zr-O bonding on the fracture surfaces and gives the lowest cleavage energy in possible crack paths. It can be summarized that the TEM/STEM investigations of the dislocations and fracture surfaces extended our knowledge and understanding of deformation and fracture in alumina in terms of atomistic mechanisms.

Acknowledgements The authors gratefully thank Takahisa Yamamoto, Teruyasu Mizuguchi, Atsutomo Nakamura, Bin Feng, Jaike Wei, Yuki Kezuka, and Akihito Ishihara for collaborative works on dislocations and grain boundary fracture in alumina. A part of this study was supported by the Elements Strategy Initiative for Structural Materials (ESISM) (Grant No. JPMXP0112101000), KAKENHI (Grant Nos. JP17H06094, JP18K13981, and 19H05788) from the Japan Society for the Promotion of Science (JSPS), "Nanotechnology Platform" (Project No. JPMXP09A20UT0080) from the Ministry of Education, Culture, Sports, Science, and Technology in Japan (MEXT), and PRESTO "Nanomechanics" by Japan Science and Technology Agency (Grant No. JPMJPR1999).

References

Bilde-Sørensen JB, Thölen AR, Gooch DJ, Groves GW (1976) Structure of $<01\bar{1}0>$ dislocation in sapphire. Philos Mag 33:877–889. https://doi.org/10.1080/14786437608221921

Bilde-Sørensen JB, Lawlor BF, Geipel T, Pirouz P, Heuer AH, Lagerlöf KPD (1996) On basal slip and basal twinning in sapphire (α-Al_2O_3)–I Basal slip revisited. Acta Mater 44:2145–2152. https://doi.org/10.1016/1359-6454(95)00264-2

Christensen A, Carter EA (1998) First-principles study of the surfaces of zirconia. Phys Rev B 58:8050–8064. https://doi.org/10.1103/PhysRevB.58.8050

Chung DH, Simmons G (1968) Pressure and temperature dependences of isotropic elastic moduli of polycrystalline alumina. J Appl Phys 39:5316–5326. https://doi.org/10.1063/1.1655961

Firestone RF, Heuer AH (1976) Creep deformation of 0° sapphire. J Am Ceram Soc 59:13–19. https://doi.org/10.1111/j.1151-2916.1976.tb09379.x

Frank FC (1951) LXXXIII. Crystal dislocations.—Elementary concepts and definitions. Philos Mag 7th Ser 42:809–819. https://doi.org/10.1080/14786445108561310

Gieske JH, Barsch GR (1968) Pressure dependence of elastic constants of single crystalline aluminum oxide. Phys Stat Sol 29:121–131. https://doi.org/10.1002/pssb.19680290113

Heuer AH, Jia CL, Lagerlöf KPD (2010) The core structure of basal dislocations in deformed sapphire (α-Al$_2$O$_3$). Science 330:1227–1231. https://doi.org/10.1126/science.1192319

Ikuhara Y, Nishimura H, Nakamura A, Matsunaga K, Yamamoto T (2003) Dislocation structures of low-angle and near-Σ3 grain boundaries in alumina bicrystals. J Am Ceram Soc 86:595–602. https://doi.org/10.1111/j.1151-2916.2003.tb03346.x

Jhon MH, Glaeser AM, Chrzan DC (2005) Computational study of stacking faults in sapphire using total energy methods. Phys Rev B 71:214101. https://doi.org/10.1103/PhysRevB.71.214101

Kenway PR (1993) Calculated stacking-fault energies in α-Al$_2$O$_3$. Philos Mag B 68:171–183. https://doi.org/10.1080/01418639308226398

Kondo S, Ishihara A, Tochigi E, Shibata N, Ikuhara Y (2019) Direct observation of atomic-scale fracture path within ceramic grain boundary core. Nature Commun 10:2112. https://doi.org/10.1038/s41467-019-10183-3

Kresse G, Furthmüller J (1996) Efficiency of ab-initio total energy calculations for metals and semiconductors using a plane-wave basis set. Comp Mater Sci 1:15–50. https://doi.org/10.1016/0927-0256(96)00008-0

Kronberg ML (1957) Plastic deformation of single crystals of sapphire: Basal slip and twinning. Acta Metall 5:507–524. https://doi.org/10.1016/0001-6160(57)90090-1

Lagerlöf KPD, Mitchell TE, Heuer AH, Rivière JP, Cadoz J, Castaing J, Phillips DS (1984) Stacking fault energy in sapphire (α-Al$_2$O$_3$). Acta Metal 32:97–105. https://doi.org/10.1016/0001-6160(84)90206-2

Lagerlöf KPD, Heuer AH, Castaing J, Rivière JP, Mitchell TE (1994) Slip and twinning in sapphire. J Am Ceram Soc 77:385–397. https://doi.org/10.1111/j.1151-2916.1994.tb07006.x

Marinopoulos AG, Elsässer C (2001) Density-functional and shell-model calculations of the energies of basal-plane stacking faults in sapphire. Philos Mag Lett 81:329–338. https://doi.org/10.1080/09500830110039984

Miao B, Kondo S, Tochigi E, Wei J, Feng B, Shibata N, Ikuhara Y (2019) The core structure of 60° mixed basal dislocation in alumina (α-Al$_2$O$_3$) introduced by in situ TEM nanoindentation. Scripta Mater 163:157–162. https://doi.org/10.1016/j.scriptamat.2019.01.011

Mitchell TE, Pletka BJ, Phillips DS, Heuer AH (1976) Climb dissociation in sapphire (α-Al$_2$O$_3$). Philos Mag 34:441–451. https://doi.org/10.1080/14786437608222034

Nakamura A, Yamamoto T, Ikuhara Y (2002) Direct observation of basal dislocation in sapphire by HRTEM. Acta Mater 50:101–108. https://doi.org/10.1016/S1359-6454(01)00318-4

Nakamura A, Matsunaga K, Yamamoto T, Ikuhara Y (2006) Multiple dissociation of grain boundary dislocations in alumina ceramic. Philos Mag 86:4657–4666. https://doi.org/10.1080/14786430600812820

Nakamura A, Tochigi E, Shibata N, Yamamoto T, Ikuhara Y (2009) Structure and configuration of boundary dislocations on low angle tilt grain boundaries in alumina. Mater Trans 50:1008–1014. https://doi.org/10.2320/matertrans.MC200821

Perdew JP, Chevary JA, Vosko SH, Jackson KA, Pederson MR, Singh DJ, Fiolhais C (1992) Atoms, molecules, solids, and surfaces: applications of the generalized gradient approximation for exchange and correlation. Phys Rev B 46:6671–6687. https://doi.org/10.1103/physrevb.46.6671

Pletka BJ, Heuer AH, Mitchell TE (1974) Dislocation structures in sapphire deformed basal slip. J Am Ceram Soc 56:136–139. https://doi.org/10.1111/j.1151-2916.1974.tb11419.x

Shibata N, Chisholm MF, Nakamura A, Pennycook SJ, Yamamoto T, Ikuhara Y (2007) Nonstoichiometric dislocation cores in α-alumina. Science 316:82–85. https://doi.org/10.1126/science.1136155

Tochigi E, Shibata N, Nakamura A, Yamamoto T, Ikuhara Y (2008) Partial dislocation configurations in a low-angle boundary in α-Al$_2$O$_3$. Acta Mater 56:2015–2021. https://doi.org/10.1016/j.actamat.2007.12.041

Tochigi E, Shibata N, Nakamura A, Mizoguchi T, Yamamoto T, Ikuhara Y (2010) Structures of dissociated <1$\bar{1}$00> dislocations and {1$\bar{1}$00} stacking faults of alumina (α-Al2O3). Acta Mater 58:208–215. https://doi.org/10.1016/j.actamat.2009.08.067

Tochigi E, Shibata N, Nakamura A, Yamamoto T, Ikuhara Y (2011) Dislocation structures in a {$\bar{1}104$}/<11$\bar{2}$0> low-angle tilt grain boundary of alumina (α-Al2O3). J Mater Sci 46:4428–4433. https://doi.org/10.1007/s10853-011-5430-y

Tochigi E, Kezuka Y, Shibata N, Nakamura A, Ikuhara Y (2012) Structure of screw dislocations in a (0001)/[0001] low-angle twist grain boundary of alumina (α-Al2O3). Acta Mater 60:1293–1299. https://doi.org/10.1016/j.actamat.2011.11.027

Tochigi E, Nakamura A, Mizoguchi T, Shibata N, Ikuhara Y (2015) Dissociation of the 1/3<$\bar{1}$101> dislocation and formation of the anion stacking fault on the basal plane in α-Al2O3. Acta Mater 91:152–161. https://doi.org/10.1016/j.actamat.2015.02.033

Tochigi E, Findlay SD, Okunishi E, Mizoguchi T, Nakamura A, Shibata N, Ikuhara Y (2016) Atomic structure characterization of stacking faults on the { 1$\bar{1}$00} plane in α-alumina by scanning transmission electron microscopy. AIP Conf Proc 1763:050003. https://doi.org/10.1063/1.496 1356

Tochigi E, Kezuka Y, Nakamura A, Nakamura A, Shibata N, Ikuhara Y (2017) Direct observation of impurity segregation at dislocation cores in an ionic crystal. Nano Lett 17:2908–2912. https://doi.org/10.1021/acs.nanolett.7b00115

Tochigi E, Mizoguchi T, Okunishi E, Nakamura A, Shibata N, Ikuhara Y (2018) Dissociation reaction of the 1/3<$\bar{1}$101> edge dislocation in α-Al2O3. J Mater Sci 53:8049–8058. https://doi.org/10.1007/s10853-018-2133-7

Sasaki T, Shibata N, Matsunaga K, Yamamoto T, Ikuhara Y (2012) Direct observation of the cleavage plane of sapphire by in-situ indentation TEM. J Ceram Soc Jpn 120:473–477. https://doi.org/10.2109/jcersj2.120.473

Chapter 8
Nanomechanical Characterization of Metallic Materials

Takahito Ohmura

8.1 Nanomechanical Characterization as an Advanced Technique

Macroscopic mechanical testing provides information on the mechanical response of materials to applied stress under various conditions. These mechanical properties are necessary for material design in engineering applications, and they are required on the scale between millimeters and meters. However, the origins in microstructures are controlled on a scale of micrometers, with the resolution in nanometers. The gap between these scales is remarkable 10^6 orders, and the gap is an extremely big hurdle in aiming at understanding the strengthening mechanism and improving the performance of structural materials.

Another big hurdle is the relation between microstructures and their mechanical properties. The macroscopic properties include the overall behavior of the material as an average quantity, but the deformation volume comprises distribution of stress and strain induced by geometrical inhomogeneity, including microstructure of materials. Although yield stress is extremely important for engineering purposes, the yielding phenomenon on a small scale, the so-called "micro-yielding," as an elemental step of macroscopic yielding, is still unrevealed. Physical models of mechanical behavior on small scales have been utilized to understand and/or predict the mechanical properties of metallic materials. The microstructures on nanometer to micrometer scale can be observed in detail with the most advanced observation apparatus, but it is still challenging to measure the mechanical behavior on the scale same as that used in the observation. As a method for describing plastic deformation quantitatively, we often

T. Ohmura (✉)

Research Center for Structural Materials, National Institute for Materials Science, Tsukuba, Japan
e-mail: ohmura.takahito@nims.go.jp

Center for Elements Strategy Initiative for Structural Materials (ESISM), Kyoto University, Sakyo-ku, Kyoto 606-8501, Japan

© The Author(s) 2022
I. Tanaka et al. (eds.), *The Plaston Concept*,
https://doi.org/10.1007/978-981-16-7715-1_8

use the dislocation theory; however, it is extremely difficult to measure the stress–strain relations directly associated with dislocation dynamics and evolution even if we can observe the behavior of individual dislocations with electron microscopes experimentally. Thus, the elucidation of the dynamic behavior of the relations, as well as removing the 10^6 order gap, is important.

Nanoindentation and micropillar testing are techniques that can be used to measure the dynamic behavior on the nanometer scale.

The nanoindentation method penetrates an indenter into a sample surface under a load of μN resolution and measures the penetration depth in nm unit, to evaluate the elastoplastic deformation of materials. The depth of the indent marks is typically below 100 nm and less than a micron horizontally. The depth is measured using a displacement gage, and then converted into the contact area by using the indenter's geometry. It can be called, in particular, depth-sensing indentation, based on the measurement principle. The details of this technique are available in the literature (Nishibori and Kinoshita 1997, 1978; Newey et al. 1982; Loubet et al. 1984; Doerner and Nix 1986; Oliver and Pharr 1992; Tsui et al. 1996; Bolshakov and Pharr 1998; Lim and Chaudhri 1999; Nix and Gao 1998; McElhaney et al. 1998; Ohmura and Tsuzaki 2007a).

Micropillar testing is a technique of uniaxial loading in compression of a columnar-shaped specimen of 100 nm–μm in diameter, typically inside the chamber of a scanning electron microscope (SEM) or a transmission electron microscope (TEM), for its in situ straining. The advanced technique is to measure the load–displacement data during staining, which provides a direct relation between microstructural evolution and mechanical behavior. In the conventional TEM in situ testing, an important point lies in observing the motion of the dislocation and microstructure evolution; however, the advanced technique is developed to measure the dynamic mechanical behavior at the same time.

In this chapter, applications of the nanomechanical characterization are described, and an elemental step and strengthening factors for the macroscopic properties are discussed.

8.2 Plasticity Initiation Analysis Through Nanoindentation Technique

A major merit of the nanoindentation technique is that it can measure less than micrometer sizes, as described above; another advantage is the fact that the underlying and fundamental behavior can be analyzed in the process of loading and unloading through consecutive measurement during the deformation. A representative example is the displacement burst phenomenon, the so-called "pop-in" event, which mostly appears in the loading process (Ohmura et al. 2001, 2002; Gerberich et al. 1996; Zbib and Bahr 2007; Ohmura and Tsuzaki 2007b, 2008; Masuda et al. 2020). Figure 8.1 shows a typical load–displacement curve for the Fe alloy, where

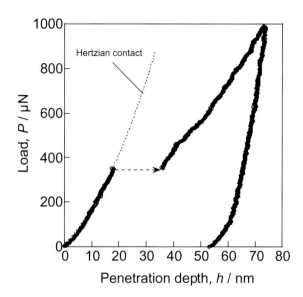

Fig. 8.1 Typical load–displacement curve for Fe alloy showing pop-in phenomenon on the loading curve, indicated by the dashed-line arrow

the pop-in phenomenon is indicated on the loading curve through the dashed-line arrow. The curve is obtained in the load-controlled mode, and the displacement burst is understood as a sudden drop in resistance to plastic deformation. When the same phenomenon is measured in the displacement-controlled mode, it appears as a sudden drop in the load (Ohmura et al. 2001). It is a drastic phenomenon to occur in a time shorter than 0.005 s, as the load–displacement data are captured 200 times per second and no point is recorded during the pop-in event.

At the beginning of the discovery of this phenomenon, it was noted that breaking a native oxide film on a sample surface might be a major reason for the event (Gerberich et al. 1996), but subsequently, it was found that the same phenomenon occurred in even noble metals such as gold, with much higher resistance to be oxidized (Ohmura et al. 2001). Additionally, the frequency of the event was higher in a sample with lower initial dislocation density, even in the same material (Ohmura et al. 2002; Zbib and Bahr 2007). Therefore, this phenomenon is considered an essential and elemental behavior of plastic deformation in materials.

Note that the pop-in phenomenon corresponds to transition from elastic to elasto-plastic deformation, and the shear stress underneath the indenter that is calculated from the applied load is close to the value of the order of the ideal strength (Gould-stone et al. 2000). This is described quantitatively, as follows, with Fig. 8.1 as an example. When we define P_c as a critical load for a pop-in event, the load–displacement curve that is lower than P_c fits very well with the dashed line of the Hertz contact model (Johnson 1985), given as

$$P = \frac{4}{3}E^* R^{\frac{1}{2}} h^{\frac{3}{2}}, \tag{8.1}$$

where R is the curvature of the indenter tip and $E*$ is the reduced modulus, given as

$$\frac{1}{E*} = \frac{1 - v_i^2}{E_i} + \frac{1 - v_s^2}{E_s},\tag{8.2}$$

where E and v are Young's modulus and Poisson's ratio, respectively, and the subscripts i and s refer to the indenter and sample, respectively. This result clearly indicates that the deformation before the pop-in event is dominated by purely elastic deformation. In addition, the maximum shear stress underneath the indenter, τ_{max}, is given as follows:

$$\tau_{max} = 0.18\left(\frac{E*}{R}\right)^{\frac{2}{3}} P^{\frac{1}{3}}.\tag{8.3}$$

When $P_c = 350$ μN, from Fig. 8.1, as P is substituted into Eq. (8.3), τ_{max} is calculated as 11.3 GPa, which is approximately 1/7th the shear modulus of 83 GPa on the order of the ideal strength. This result indicates that the pop-in phenomenon corresponds to the plasticity initiation by dislocation nucleation from a region with no pre-existing lattice defect. On the other hand, a previous study conducted using the TEM in situ straining technique revealed that, in pure Al, dislocations are nucleated prior to the pop-in event (Minor et al. 2006). This indicates that some other processes, such as dislocation multiplication, occur subsequent to the dislocation nucleation to occur a pop-in event.

As an example of how the multiplication process of dislocations is related to the mechanical behavior, the results of TEM in situ compression deformation are shown. The specimen is pillar shaped for the compression test of Fe -3 wt% Si single crystal, with the compression axis being parallel to the <110> axis. By measuring the load–displacement relation during compression deformation simultaneously, the mechanical response and dislocation structure change can be synchronized. Figure 8.2 is a snapshot extracted from the recorded movie and shows the change in the dislocation structure during deformation (Ohmura et al. 2012). Figure 8.2a shows the snapshot right before the pop-in event, and (b) corresponds to that immediately after the pop-in event. In (b), an increase in the dislocation density is clearly observed, indicating that a multiplication of dislocations occurs during the pop-in event. The time difference between (a) and (b) is 1/30 s, indicating that the dislocation multiplication and propagation occur within a very short period. Figure 8.2c shows the load–displacement curve measured during TEM in situ straining, where the dashed arrow indicates a sudden drop in the load at the moment of pop-in. As this experiment was conducted in the displacement-controlled mode, the behavior is different from that of the displacement burst, which appears in the load-controlled mode; however, both of them show a drastic decrease in the deformation resistance, and thus can be regarded as essentially the same phenomena. This result indicates that dislocation multiplication is an important elementary process in deformation behavior, where

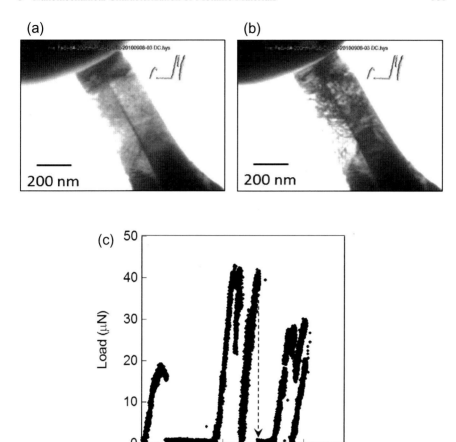

Fig. 8.2 Snapshot extracted from the recorded movie, showing changes in the dislocation structure during deformation. **a** Shows the snapshot right before the pop-in and **b** corresponds to the snapshot immediately after the pop-in. **c** Load–displacement curve measured during TEM in situ straining (Ohmura et al. 2012). Reprinted with permission from [T. Ohmura, L. Zhang, K. Sekido and K. Tsuzaki: J. Mater. Res., 27 (2012), 1742–1749.] Copyright (2012) by Cambridge University Press

the plastic strain increases rapidly. The relation between the evolution of dislocation structure and plastic strain is discussed further in Sect. 8.4.

To further understand the factors governing the initiation behavior of local plastic deformation, a systematic analysis was performed using various single crystals with a variety of materials with different crystal structures. All vertical directions of the sample surface were oriented to <001> . Figure 8.3 shows the relation between the maximum shear stress τ_{max}, calculated from the pop-in load P_c using Eq. (8.3), and the stiffness modulus G, converted from Young's modulus calculated from the

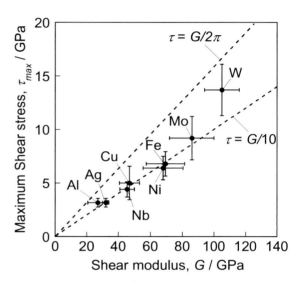

Fig. 8.3 Relationship between the maximum shear stress τ_{max} calculated from the pop-in load P_c using Eq. (8.3) and the stiffness modulus G converted from the Young's modulus calculated from the unloading curve (Ohmura et al. 2012). Reprinted with permission from [T. Ohmura, L. Zhang, K. Sekido and K. Tsuzaki: J. Mater. Res., 27 (2012), 1742–1749.] Copyright (2012) by Cambridge University Press

unloading curve (Ohmura et al. 2012). This relation was linear for all measured materials, and the coefficient was found to be close to $1/2\pi$.

On the other hand, one of the models in which the frictional stress of the perfect crystal on the slip plane was formulated is given as

$$\tau = \frac{b}{h}\frac{G}{2\pi}\sin\left(\frac{2\pi x}{b}\right), \tag{8.4}$$

where b is the magnitude of the Burgers vector, h is the distance between the slip planes, and x is the relative displacement in the slip direction. The maximum stress obtained by approximating b to h is $G/2\pi$ when $x = b/4$. This value is very close to that obtained experimentally in Fig. 8.3. This result strongly evidences that the stress level at which the pop-in behavior appears is close to the ideal strength regardless of the crystal structure, indicating that the critical stress strongly depends on the local shear modulus.

8.3 Effect of Lattice Defects Including Grain Boundaries, Solid-Solution Elements, and Initial Dislocation Density on the Plasticity Initiation Behavior

8.3.1 Grain Boundary

The model of grain refinement strengthening is often used to discuss the strengthening mechanism induced by grain boundaries. Grain refinement strengthening is often described by the following Hall–Petch relation (Hall 1951; Petch 1953), formulated from the experimental results as

$$\sigma = \sigma_0 + kd^{-1/2}. \tag{8.5}$$

Here, σ refers to the flow stress σ_0 is a constant, k is the locking parameter, and d denotes the grain size. The pile-up model (Hall 1951; Petch 1953) and dislocation source model (Li 1963) are shown as the mechanisms for grain boundary strengthening. The former is understood as a function of resistance against dislocation motion, while the latter as a function of enhancing the dislocation generation, which seem to contradict each other at first glance. To verify these models, it is effective to directly capture the interaction between a single grain boundary and dislocation. However, in previous studies, only the microstructural observation by TEM (Hauser and Chalmers 1961; Carrington and McLean 1965; Shen et al. 1988; Kurzydlowski et al. 1984; Lee et al. 1990) has been conducted, and not the quantitative evaluation of the mechanical behavior. To address this problem, the authors performed nanoindentation in the vicinity of a single grain boundary and verified these two models from the viewpoint of mechanical behavior as described below.

The sample was Ti-added interstitial free (IF) steel, with an average grain size of several 100 μm. Details of the experimental method are shown elsewhere (Ohmura et al. 2005). The indented positions were set in two ways: just above the grain boundary and within the grain far from the grain boundary. Figure 8.4 shows an example of the load–displacement curves obtained from the nanoindentation measurements. The pop-in event clearly appeared on both the grain boundary (open circle) and grain interior (dot) on the loading curves. Figure 8.5 shows the relation between the critical load P_c and pop-in depth Δh. An example SPM image of the sample surface after the nanoindentation is inserted in the bottom right of the figure to

Fig. 8.4 Example of load–displacement curves obtained from nanoindentation measurements. Pop-in clearly appears on both the grain boundary (open circle) and grain interior (dot) on the loading curves (Ohmura et al. 2005). Reprinted with permission from [T. Ohmura, K. Tsuzaki and F. Yin: Mater. Trans., 46 (2005), 2026–2029.] Copyright (2005) by The Japan Institute of Metals and Materials

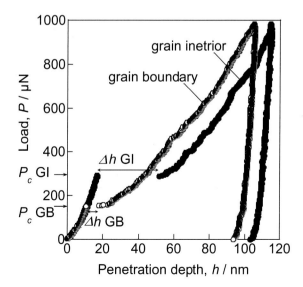

Fig. 8.5 Relationship between the critical load P_c and the pop-in depth Δh. An example of SPM image of the sample surface after the nanoindentation is inserted in the bottom right on the figure to show that the indent mark is formed certainly just above the grain boundary (Ohmura et al. 2005). Reprinted with permission from [T. Ohmura, K. Tsuzaki and F. Yin: Mater. Trans., 46 (2005), 2026–2029.] Copyright (2005) by The Japan Institute of Metals and Materials

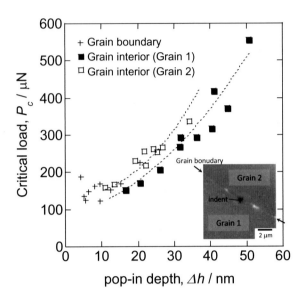

show that the indent mark is certainly formed just above the grain boundary. The two grains forming the grain boundary are named as grains 1 and 2, as shown in the figure, and the data of each grain's interior are plotted separately. On the grain boundary, P_c has relatively lower values of 100–200 μN, whereas in the grain interior, it is dispersed up to approximately 600 μN. The results suggest that grain boundaries act as effective dislocation sources for enhancing the dislocation emission for plasticity initiation. One of the characteristics of the grain interior data is that a higher P_c leads to a higher Δh. This can be understood in the following relation, derived from the model (Shibutani and Koyama 2004) in which plastic deformation upon the pop-in event is initiated by the prismatic loop dislocation generation (Ohmura et al. 2005):

$$P_c = \left(\frac{1}{0.18}\right)^3 \left(\frac{R}{E^*}\right)^2 \left(\frac{2G}{a}\Delta h + \mu\gamma\right)^3, \tag{8.6}$$

where a indicates the horizontal size of the indent and γ denotes the elastic strain remaining after the pop-in. Equation (8.6), drawn by the broken line in the figure, fits well with the experimental data of the relation between P_c and Δh, and the higher the value of P_c, that is, the greater the accumulated elastic strain energy, the greater is the plastic strain at pop-in.

The reason why P_c is dispersed in the range of 100–600 μN is discussed below. The IF steel used in this experiment may have a higher dislocation source density before the indentation experiment as compared to the single-crystal sample shown in the previous section. When the initial dislocation source density is low, the indentation-induced stress at the pre-existing dislocation source does not reach the critical stress for dislocation source activation, and thus the plastic deformation must be initiated through the generation of dislocations from the perfect crystal region. In contrast, when the dislocation source density is high, the plastic deformation starts with the activation of the pre-existing dislocation source without the nucleation of dislocation. The reason why P_c is distributed widely when the initial dislocation source density is high is, first, that the stress applied to the dislocation source varies depending on the position, because the stress field introduced by the indentation has a distribution. The maximum shear stress shown in Eq. (8.3) is the stress generated at a point just below the indenter, and it is rather rare that the position of the dislocation source coincides with that point and is further away from this point with a lower applied stress. Second, even if the dislocation source is located at the same position with the maximum shear stress, the critical stress required for the activation may depend on individual cases. Assuming, for example, a Frank Reed (FR) source as a dislocation source, the activating critical stress is Gb/l (l is the length of the FR source), where l may assume various values.

On the other hand, in the nanoindentation measurement, the hardness can be calculated from the indentation depth corresponding to the maximum load, as in the conventional method, and is determined as 2.8 ± 0.16 and 2.2 ± 0.05 GPa at the grain boundary and in the grain interior, respectively. The deformation resistance at the grain boundary is approximately 30% higher than that in the grain interior. This is interesting in contrast to the behavior described in Fig. 8.5, where the plastic deformation is initiated at the grain boundary at a lower stress than that in the grain interior. That is, while a single grain boundary initiates plastic deformation by acting as an effective dislocation source, when dislocation sources other than grain boundaries are activated in the further plastic deformation, the single grain boundary acts as a resistance against the sliding motion of dislocations moving toward the grain boundary, which indicates the remarkable contribution of the single grain boundary to strengthening in a certain strain region.

8.3.2 Solid Solution Element

Figure 8.6 shows typical load–displacement curves obtained by nanoindentation for Fe–C binary alloys with various carbon contents (Nakano and Ohmura 2020). In all loading processes, a displacement burst, indicated by dashed arrows, i.e., pop-in, occurred. As shown in Fig. 8.6, the critical load P_c at which pop-in occurs increases with the concentration of in-solution carbon.

Figure 8.7 shows the relation between the τ_{max} obtained by substituting the P_c value into Eq. (8.3), shown in Fig. 8.6, and the carbon concentration. The plots are averages and the error bars are standard deviations (SDs). As the carbon concentration increases, both τ_{max} and the SD increase.

To clarify the variation in the deviation, the probability distribution of P_c for each sample is shown in Fig. 8.8. The distribution of P_c is Gaussian-like at 0 C and 3 C, with a peak at around 350 µN. On the other hand, at a higher carbon concentration, the peak height around 350 µN decreases and another peak appears at the higher

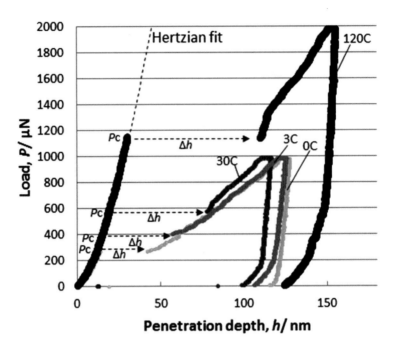

Fig. 8.6 Typical load–displacement curves for all samples (Nakano and Ohmura 2020).]. Reprinted with permission from [K. Nakano and T. Ohmura, J. Iron and Steel Inst. Japan, 106 (2020), 82–91.] Copyright (2020) The Iron and Steel Institute of Japan

Fig. 8.7 Relationship between solute carbon concentration and critical pop-in load Pc (Nakano and Ohmura 2020). Reprinted with permission from [K. Nakano and T. Ohmura, J. Iron and Steel Inst. Japan, 106 (2020), 82–91.] Copyright (2020) The Iron and Steel Institute of Japan

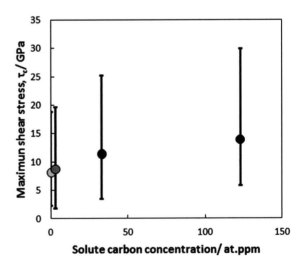

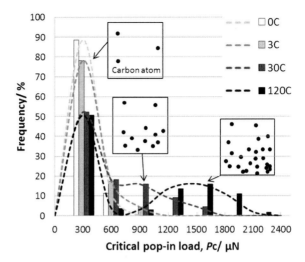

Fig. 8.8 Frequency of the critical pop-in load (Nakano and Ohmura 2020). Reprinted with permission from [K. Nakano and T. Ohmura, J. Iron and Steel Inst. Japan, 106 (2020), 82–91.] Copyright (2020) The Iron and Steel Institute of Japan

load over 500 μN. In addition, the peak position shifts to a higher load and the peak width widens for 120 C, compared to the case of 30 C. The pop-in phenomenon is controlled by the thermal activation process, because both peaks are Gaussian-distributed regardless of the position of the peak. Accordingly, the thermal activation process seems to be dominant for the pop-in generation, even if the solid-solution carbon atom is related.

The effect of solid-solution carbon on the pop-in phenomenon is discussed based on the mechanism of dislocation nucleation. The pop-in phenomenon corresponds to the onset of plastic deformation, as mentioned above. We previously (Zhang and Ohmura 2014) demonstrated experimentally that the pop-in phenomenon corresponds to the nucleation of dislocations from a defect-free region (see details in Sect. 8.4). Lorentz et al. (2003) concluded that dislocation generation is a homogeneous nucleation of the shear dislocation loop, because the experimentally measured shear stress at pop-in agrees well with the ideal strength. On the other hand, Schuh et al. (2005) found the activation volume from the probability distribution function of the pop-in stress and indicated that the inhomogeneous nucleation dominates the event because the activation volume is very small below 1.0 b^3 (b is the magnitude of Burgers vector). As the occurrence of the shear dislocation loop is considered an elementary process in both cases, Sato et al. (2019) modeled this process using the molecular dynamic simulation at finite temperature and showed that the temperature dependence of τ_{max} agreed well with the experimental results of W and Fe. Thus, experiments and atomic simulations show that dislocation nucleation is necessary for pop-in formation, and the thermal activation process is considered dominant even in the presence of solid-solution carbon atoms. Thus, these carbon atoms increase τ_{max} to resist this nucleation process. A detailed model of this mechanism is described later.

The reason why the frequency distribution of P_c varies with the carbon concentration is discussed subsequently. As shown in Fig. 8.8, as the carbon concentration increases, the peak around 350 μN remains constant, while the other peak position shifts to a higher load. This trend suggests a nonuniform carbon distribution. The peak at 350 μN corresponds to the behavior under the carbon-free condition because the peak appears even in the 0 C sample. If the spatial distribution of carbon atoms remains uniform after the in-solution carbon concentration increases, only the average value is expected to increase while the distribution shape remains a single peak. Therefore, the multiple peaks suggest that different mechanisms dominate the pop-in behavior. As the peak position at 350 μN is constant regardless of the carbon concentration and the peak is highest in the 0 C sample, the mechanism is governed by the same resistance to dislocation nucleation, where in-solution carbon atoms are hardly involved. On the other hand, the peak at the higher load positions that appears after the carbon addition is considered to be caused by the interaction between single or multiple in-solution carbon atoms and dislocations with higher resistance to dislocation nucleation. The reason why the peak position shifts to the high load side with an increase in the carbon concentration is attributed to the shortening of the distance between the carbon atoms in solution. These are considered from the dislocation nucleation model developed by Sato et al. (2019), as follows. Assuming that the elementary process of the pop-in phenomenon is the nucleation of the shear dislocation loop, the applied force is balanced with a line-tension force under the condition of a loop curvature smaller than the critical size. Therefore, when the diameter of the shear dislocation loop is smaller than the critical size, the loop disappears upon unloading. To reach and overcome the critical size for a stable growth of shear dislocation loops, it is necessary to further increase the applied force or weaken the line tension through thermal fluctuation. That is, it is necessary to exceed the critical size of the dislocation loop in order for the dislocation to nucleate and pop-in to occur. Under a certain external force, dislocation loops of various sizes are generated by the thermal activation process, while those below the critical size disappear. The higher the applied stress, the lower is the activation energy required for dislocation nucleation and the smaller is the critical radius. The above model indicates that a higher stress is required at the same temperature to reach the critical size under the effect of in-solution carbon atoms, because carbon atoms generate resistance to the growth of dislocation loops by pinning the migration of dislocation lines. Therefore, the peak of the higher load side, which appears when the carbon concentration increases, appears because single or multiple solid-solution carbon atoms act as a large resistance to dislocation motion. As the solid-solution carbon concentration increases, the distance between the in-solution carbon atoms decreases and some atoms may form a cluster (Ushioda et al. 2019). It is considered that the peak position shifts to the higher load side with an increase in the carbon concentration, cluster number density, and cluster size.

The reason why the distribution width on the high load side increases with the carbon concentration needs to be discussed. This is synonymous with the increase in the error bar shown in Fig. 8.7 and is considered to be an essential result of the distribution of in-solution carbon atoms rather than the measurement error. When the carbon concentration increases, the average distance between the carbon atoms decreases and the formation frequency of the local aggregates, such as a cluster, increases. There are various cluster sizes under this condition, and the spatial distribution of number density arises. The reason why the distribution width of the high load side expands is attributed to the fact that the distribution of number density and cluster size increases with the carbon concentration. On the other hand, the distribution width on the low load side does not change even when the carbon concentration increases. As described above, this peak is attributed to the mechanism in which carbon atoms are not involved, and therefore there might be a region in which carbon atoms do not exist depending on the measurement position, even if the sample includes nominal carbon atoms. The volume of this region is expected to be more than that of the plastic region formed underneath the indenter. It is reported that the diameter is approximately 10 times the indentation depth when the plastic region under the indenter is assumed to be hemispherical (Itokazu and Murakami 1993). As the indentation depth corresponding to the peak on the low load side is 10–20 nm, the corresponding diameter of the plastic region is estimated to be 0.1–0.2 μN. That is, the region with no carbon atoms is estimated to be larger than this size, and this region is considered to be randomly dispersed in the sample. Based on the model, as shown in the schematic diagram in Fig. 8.8, as the nominal carbon concentration increases, the local concentration increases in the region where solid-solution carbon atoms exist, while there are regions where almost no atoms exist, which leads to a significant inhomogeneity in the distribution of these atoms.

8.3.3 Initial Dislocation Density

Pre-existing lattice defects, including dislocations, affect the behavior of plasticity initiation underneath an indenter. The STEM micrographs for ultra-low carbon (ULC) and IF steels with different dislocation densities are shown in Fig. 8.9 (Sekido et al. 2012). The recrystallized samples are tensile-deformed up to approximately 40% strain at room temperature to get a high dislocation density of 10^{14} m^{-2}. For the other specimens, the recrystallized samples are further annealed at 1123 K for 7.2 ks, and then held at 973 K for 1.8 ks, followed by cooling to room temperature in air, obtaining a low dislocation density of 10^{11} m^{-2}. Figure 8.10 shows typical load–displacement curves for (a) the specimens after tensile deformation and (b) the full-annealed one in both ULC and IF steels. Pop-in phenomenon appears clearly in the low-dislocation-density steels in Fig. 8.10a, and the critical load for the pop-in in ULC is higher than that in the IF. On the other hand, no clear pop-in is observed in steels with high dislocation density shown in Fig. 8.10b. Even though the pop-in phenomenon is not clear in Fig. 8.10b, the critical pop-in load P_c can be found using the Hertz contact curve of Eq. (8.1) by a deviation from the broken line. Compared to Fig. 8.10a, b, the P_c values are extremely low and the effect of solid-solution elements is unclear in (b), with a higher dislocation density. Figure 8.11 a–d shows the plots of P_c versus Δh for IF and ULC with low and high dislocation densities. The following

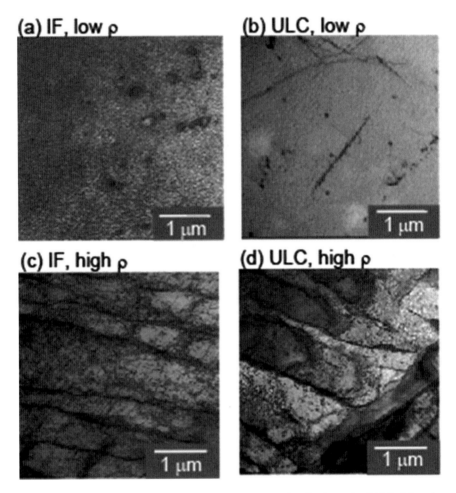

Fig. 8.9 STEM images for IF and ULC with **a, b** low dislocation density and **c, d** high dislocation density (Sekido et al. 2012). Reprinted with permission from [K. Sekido, T. Ohmura, T. Hara and K. Tsuzaki: Mater. Trans., 53 (2012), 907–912.] Copyright (2012) by The Japan Institute of Metals and Materials

three points can be gotten from the figures. First, the average P_c in the ULC steel is higher than that in the IF steel with low dislocation density. Second, the average P_c in the high-dislocation-density samples is lower than that in the low-dislocation-density materials. Third, the average P_c in the ULC steel is almost the same as that in the IF steel with high dislocation density. Leipner et al. (2001) described the critical stress τ_n for the dislocation nucleation in GaAs under indentation-induced stress field. The equation is given as

$$\tau_n = \frac{Gb}{\pi e^3 r_0} \frac{2 - \nu}{1 - \nu}, \tag{8.7}$$

Fig. 8.10 Typical
load–displacement curves
for IF and ULC with **a** low
dislocation density and **b**
high dislocation density
(Sekido et al. 2012).
Reprinted with permission
from [K. Sekido, T. Ohmura,
T. Hara and K. Tsuzaki:
Mater. Trans., 53 (2012),
907–912.] Copyright (2012)
by The Japan Institute of
Metals and Materials

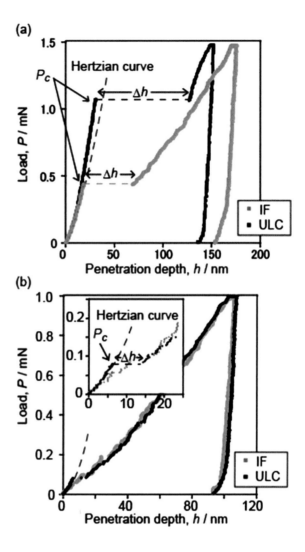

where e is the Euler number and r_0 is the cutoff radius at the dislocation core. We
obtain $\tau_n \approx 9.7$ GPa using the typical values of $r_0 = b/3$, $b = 0.29$ nm, $G = 83$ GPa,
and $v = 0.3$ for ferrite. Meanwhile, the maximum shear stress τ_{max} underneath the
indenter can also be determined from Eq. (8.3) to be approximately 13 and 18.5 GPa
for IF and ULC steels, respectively. In the high-dislocation-density materials, τ_{max} is
lower than τ_n, suggesting that the plasticity initiation is dominated by not dislocation
nucleation but another mechanism with a lower critical stress. In high-dislocation-
density materials, the microstructure can include numerous dislocation sources that
have been generated by lattice defects reactions during the tensile deformation, and
some dislocation sources may be activated at a lower applied stress than the critical
shear stress τ_c for the indentation-induced dislocation source, and/or pre-existing

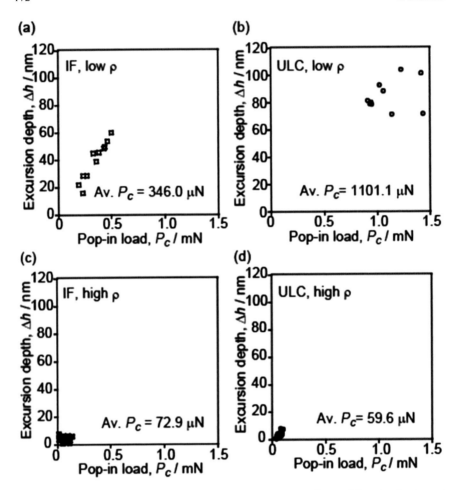

Fig. 8.11 Relationship between P_c and Dh for IF and ULC with **a, b** low dislocation density and **c, d** high dislocation density (Sekido et al. 2012). Reprinted with permission from [K. Sekido, T. Ohmura, T. Hara and K. Tsuzaki: Mater. Trans., 53 (2012), 907–912.] Copyright (2012) by The Japan Institute of Metals and Materials

dislocations may start to move at a lower stress than that of dislocation nucleation. Accordingly, the plasticity initiation under an indentation-induced applied stress is presumably dominated by the multiplication and/or inception of glide motion of a pre-existing dislocation underneath the indenter. Consequently, the plastic deformation is initiated at a lower load.

There is no significant difference in P_c between the IF steel in Fig. 8.11c and the ULC steel in Fig. 8.11d, indicating that interstitial carbon has no effect on the pop-in event in high-dislocation-density materials. The thermal activation process of the dislocation motion is discussed subsequently to determine the effect of interstitial carbon on the pop-in event on an experimental approach. The passing mechanism of dislocation on the interstitial carbon should be a thermal activation process because interstitial carbon is a short-range obstacle for dislocation glide motion. Therefore, the pop-in behavior could be affected by the interstitial carbon in ULC steel, and P_c should show an indentation rate dependence (Sekido et al. 2011a). Figure 8.12a, b shows the loading rate dependence of P_c with low- and high-dislocation-density materials, respectively. In low-dislocation-density materials, the P_c in ULC steel shows a clear indentation rate dependence, while in high-dislocation-density materials, no dependence is shown in IF and ULC steels. These results indicate that there

Fig. 8.12 Indentation rate dependence for IF and ULC with **a** low dislocation density and **b** high dislocation density (Sekido et al. 2012). Reprinted with permission from [K. Sekido, T. Ohmura, T. Hara and K. Tsuzaki: Mater. Trans., 53 (2012), 907–912.] Copyright (2012) by The Japan Institute of Metals and Materials

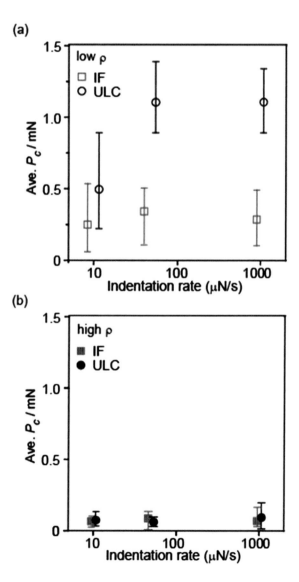

is no effect of interstitial carbon on the pop-in event of the plasticity initiation for high-dislocation-density materials.

Next, we discuss two reasons why carbon does not have any effect on the ULC with high dislocation density under the occurrence of dislocation multiplication.

The first reason is that the dislocation with no carbon pinning could be the dominating mechanism. Britton et al. (2009) discussed the relation between the amount of carbon and the dislocation density for a pop-in event, using the Fe—0.01 wt%C polycrystal (bcc) based on Cottrell and Bilby's model (1949). This model explains the effect of carbon content, ω (wt%), on the stress–strain curve of steels with different dislocation densities ρ (m^{-2}). A yield drop occurs on the stress–strain curve when

$$\omega/\rho \geq 10^{-18} \left(\text{wt\%m}^2 \right). \tag{8.8}$$

In contrast, no yield drop appears when

$$\omega/\rho \geq 10^{-19} \left(\text{wt\%m}^2 \right). \tag{8.9}$$

Thus, they concluded that pop-ins do not occur in the sample with many dislocations, as there are not enough carbon atoms to pin all dislocations, which is analogous to the case of the ULC with high dislocation density in our study. In the ULC steel, the carbon content is 0.0038 wt%, and dislocation densities are 10^{11} m^{-2} and 10^{14} m^{-2} for lower and higher samples, respectively. ω/ρ is estimated to be 10^{-14}(wt%m^2) for the low dislocation density, which satisfies Eq. (8.8), resulting in an occurrence of pop-in. On the other hand, ω/ρ is calculated as 10^{-17}(wt%m^2) for high-dislocation-density sample, which is two orders higher with the critical value in Eq. (8.9). However, the carbon content can be overestimated in the grain interior because the carbon atoms tend to segregate to the grain boundaries and therefore the actual carbon content in the grain interior can be lower than the nominal value in the whole sample. Additionally, the dislocation density can be underestimated as we can measure it only in the interior of the dislocation cells and cannot count the dislocations on the cell walls. Therefore, the real value of ω/ρ can be much lower than the estimated, corresponding to the case of Eq. (8.9). Consequently, we can presume that a part of the pre-existing dislocations is free from pinning by carbon and can move in the same manner as the dislocations in the IF as many pre-existing dislocation and sources exist underneath the indenter.

The other reason is that the critical stress required for dislocation source activation is more dominant than unpinning from carbon. We estimate the balance between the carbon contents and the dislocation density for the ULC steel with high dislocation density. In this estimation, all carbon atoms are assumed to exist in the grain interior with no segregation or precipitation. The number of carbon atoms per unit volume, $N^c{}_v$, in the ULC steel that is estimated from the carbon composition is as follows. The carbon content in the ULC is 0.0177 at%; thus, the average spacing of carbon is estimated to be approximately 4 nm, and $N^c{}_v$ is calculated to be 1.6×10^{25} m^{-3}. On the other hand, the number of carbon atoms segregating on a dislocation, $N^d{}_v$, is calculated from the dislocation density. We assume the spacing of the carbon to be

0.29 nm, which is the nearest neighbor of the octahedral site. $N^d{}_v$ is obtained as the dislocation density (10^{14} m^{-2}) divided by the spacing of carbon atoms (0.29 nm) to be 3.4×10^{23} m^{-3}. Based on the estimations, $N^c{}_v$ is much larger than $N^d{}_v$, indicating that the carbon content is high enough to pin all dislocations. Thus, another possibility should be considered. On the other hand, in the line-tension model of dislocation, the critical stress τ_p required for the dislocation multiplication from FR-type source is expressed as

$$\tau_p = \frac{Gb}{l_p},\qquad(8.10)$$

where l_p is a FR length. τ_p is dominant if it is greater than the stress required for unpinning from the carbon. Even though it is not easy to estimate the stress required for unpinning from the carbon in an individual dislocation, the yield stress given by the tensile test is approximately 300 MPa (Sekido et al. 2011a); hence, a high probability is given for dislocation glide motion at this stress level for getting a certain plastic strain in the initiation of deformation. On the other hand, the τ_{max} calculated from P_c in Eq. (8.3) is approximately 8 GPa. Since there is a stress distribution underneath the indenter and the position of the activated dislocation source can be far from the position of τ_{max}, the actual τ_p might be lower than 8 GPa. However, the τ_p value can still be much higher than the macro-yield stress level. Thus, the critical stress τ_p for the activation of the pre-existing dislocation and dislocation source that is induced by tensile deformation is dominant for the pop-in event and unpinning from the carbon has no effect on the ULC with high dislocation density. In this case, P_c is associated with the τ_p in Eq. (8.10); hence, in-solution carbon is not related to P_c. This result is similar to the low-average P_c at grain boundaries, indicating that the presence of the dislocation source is related to the initiation behavior of plastic deformation.

8.4 Initiation and Subsequent Behavior of Plastic Deformation

8.4.1 Sample Size Effect and Elementary Process

One factor that determines the critical stress τ_y for the onset of plastic deformation on a slip plane is the Peierls potential, which changes the self-energy of dislocations due to the periodicity of the crystal structure. The parameter τ_y indicates the stress required to cross one of the peaks of the Peierls potential (Peierls stress) without thermal assistance. It has been shown experimentally for various materials that τ_y is expressed in the following form by using the shear modulus G, the spacing between lattice planes h, and the magnitude of Burgers vector b (Takeuchi and Suzuki 1989;

Suzuki and Takeuchi 1989).

$$\tau_y = \alpha G \exp\left(-\beta \frac{h}{b}\right),$$ (8.11)

where α and β are constants. As τ_y is normalized by G and τ_y / G depends only on the value of h/b, the critical shear stress is almost determined by the crystal structure and lattice constant. As described above, in the case of a high-purity crystal, especially a crystalline material other than a metal and low-temperature deformation of bcc metal, the yield stress is determined by the intrinsic factor of the crystal for dislocation glide motion.

The above interpretation is a model for understanding the macroscopic yield phenomenon of a single crystal by the motion of a single dislocation. In other words, the model assumes that a mobile dislocation already exists in the crystal and that its motion governs the plasticity.

This raises one question. Where does the mobile dislocation originate in the crystal before yielding? Is it the so-called grown-in dislocation, or is it a dislocation formed or grown during deformation?

The high strength of the whisker was originally attributed to its defect-free nature, but it is understood that the strength strongly depends on the sample size and that the thinner the crystal, the lesser is the probability that a longer dislocation source exists in the sample (Brenner 1956). More recently, Uchic et al. (2004), in a systematic study of cylindrical single-crystal samples of sizes ranging from 0.5 to several 10 μm, showed that the yield stress measured in compression tests increased with the decreasing cylinder diameter. The reason for this is attributed to the fact that the smaller the sample size, the lower is the number of initial mobile dislocations involved. Note that the yield stress depends on the sample size, that is, the critical shear stress of a slip plane depends on the number of dislocations or the number of dislocation sources included in the stress field. This may provide a new insight into the strengthening mechanism in grain refinement strengthening, for example. The conventional mechanism model of grain refinement strengthening is grain boundary strengthening, which is an obstacle to the glide motion of dislocations at grain boundaries as an elementary process. Besides the conventional model, it is also necessary to consider that the yield stress in the grain interior increases due to the grain size effect. The authors demonstrated that the hardness of the grain interior, as well as the macro-hardness for ultrafine-grained materials of IF steel and pure Al, increases according to the grain size (Ohmura et al. 2004a; Zhang et al. 2009). The fact that the yield stress in the grain interior depends on the grain size indicates that the σ_0 term in Eq. (8.5), which has been considered to be the material constant regardless of the grain size, does not have a constant value and contains important suggestions concerning the strengthening mechanism.

Another important point in the dependence of strength on the sample size is that not only the mobile dislocation density but also the dislocation source density is related to the mechanical behavior. In dislocation theory, when the dislocation density ρ at a certain instant is constant, the strain rate $\dot{\gamma}$ is expressed by a model governed by the average mobility of dislocations $\bar{v}$ and b, as follows:

$$\dot{\gamma} = \rho b \, \bar{v}. \tag{8.12}$$

This model is useful for explaining the thermal activation process of deformation based on the temperature dependence of the dislocation mobility by the finite velocity caused by viscous motion. In the discussion of the macroscopic plastic strain, a statistical mechanical approach is possible because it can be regarded as the average velocity of many dislocations or that of one dislocation during its long-range motion. The analysis of the general thermal activation process is often based on this model.

On the other hand, in the case of flight motion in which the generated or grown dislocations move at a constant distance instantaneously, the strain rate $\dot{\gamma}$ is given by

$$\dot{\gamma} = \dot{\rho} b \bar{x}, \tag{8.13}$$

where $\dot{\rho}$ is the increase rate of dislocation density and $\bar{x}$ is the average travel distance of the dislocation glide motion. In this case, $\bar{x}$ is considered to be the distance to a mechanical equilibrium position or to the surface, which can be considered constant in a short time. Therefore, the strain rate is controlled by the growth rate of dislocations $\dot{\rho}$ at a moment. This model seems to be close to reality as an elementary process of plastic deformation, if the material deformation is modeled locally and/or from the short time viewpoint. In fact, as clearly observed in Fig. 8.2, the dislocation density rapidly increases in a very short time and the generated dislocation immediately moves a certain distance. For the bulk material, the following results are obtained on the dislocation structure introduced upon pop-in Zhang and Ohmura (2014). Figure 8.13 shows (a) the SPM image of indentation marks on the sample surface, (b) the corresponding load–displacement curves, and (c) the cross-sectional TEM images of the dislocation structure just below the indent marks. As shown in (a), even though the same peak load is applied to 3×8 regularly arrayed positions, no indentation is formed in some cases. For example, the load–displacement curves of 1–3 in (b) correspond to their indent marks in the bottom row on the SPM image in (a). In case of indent 2, no indentation is formed in (a), and the load–displacement curve in (b) shows a complete elastic deformation in which the loading and unloading curves overlap, and no dislocation is observed in the TEM image in (c). On the other hand, in case of indent 3, the unloading starts immediately after pop-in according to the load–displacement behavior in (b), and therefore almost all dislocations observed in (c) are introduced at the moment of pop-in. Compared to case 1, in which certain plastic deformation progresses after pop-in, dislocation structures in cases 1 and 3 are not much different in terms of distribution range and many complicated dislocation lines. A comparison of these three cases indicates that several dislocations are generated at the instant of pop-in and move to a certain distance, and then the deformation progresses while the dislocation structure develops gradually in the subsequent deformation. That is, the growth-dominated deformation of Eq. (8.13) progresses at the

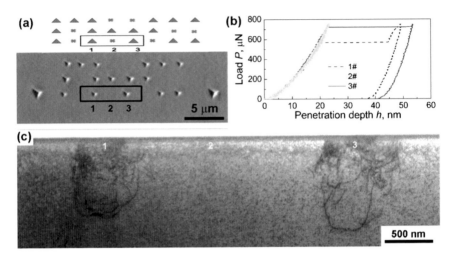

Fig. 8.13 **a** SPM image of indentation marks on the sample surface, **b** the corresponding load–displacement curves, and **c** the cross-sectional TEM images of the dislocation structure just below the indent marks (Zhang and Ohmura 2014). Reprinted with permission from [L. Zhang and T. Ohmura: Phys. Rev. Lett., 112 (2014), 145,504.] Copyright (2014) by The American Physical Society

instant of pop-in, and thereafter the mobility-dominated deformation of Eq. (8.12) progresses.

Although the relation between the two models is controversial, both are very important in discussing the mechanism of plastic deformation. As it was conventionally difficult to capture individual dislocation motions directly, an average handling had to be conducted. The model expressed in Eq. (8.12) is suitable for discussing the macroscopic behavior. On the other hand, the recent advanced observation and analysis technologies, including TEM in situ straining, have made it possible to approach the behavior on the scales of micron or less, as well as more precise elementary process analysis (Ohmura et al. 2004b; Zhang et al. 2011; Carpeno et al. 2015; Hsieh et al. 2016; Chung et al. 2018; Onose et al. 2019; Kim et al. 2012). It is necessary to verify how the conventional knowledge matches the new knowledge by the advanced research approach or whether it does not match without sticking to the complexes.

8.4.2 Dislocation Mobility and Mechanical Behavior in Bcc Crystal Structures

Note that the $1/2 <111>$ screw dislocations in bcc crystals exhibit a large Peierls force due to the specificity of the atomic arrangement and dislocation core structure (Hirsch 1968; Vitek 1974; Suzuki 1968; Takeuchi 1981; Edagawa et al. 1997), and they are considered to have significantly lower mobility than edge dislocations. It has been experimentally observed that linear screw dislocations are dominant in dislocation structures after deformation in bcc single crystals. This behavior is understood as the reason for the anomalous properties in that the yield stress of bcc crystals is higher than that of fcc crystals in the low-temperature range, the temperature dependence of the yield stress is large, and CRSS shows plastic anisotropy depending on the shear direction (failure of Schmid's law).

In general, the relation between dislocation mobility and applied stress is expressed by the following equation, as dealt with in Johnston–Gilman's theory (Johnston and Gillman 1959; Johnston 1962):

$$\bar{v} \propto \tau^m. \tag{8.14}$$

The stress exponent m varies depending on the crystal and is relatively small in the case of a semiconductor or an ionic crystal and large in the case of a metal. From Eqs. (8.12) and (8.14), the relation between strain rate $\dot{\gamma}$ and applied stress τ is expressed as follows:

$$\dot{\gamma} \propto \rho b \tau^m. \tag{8.15}$$

Therefore, under the condition that m is small, the strain rate dependence of the applied stress increases and the mobility of the dislocation dominates the yield stress. Based on this model, the reasons why the temperature dependence of the yield stress of bcc crystals is high can be understood by the low mobility of screw dislocations.

The relation between the dislocation mobility and applied stress is directly observed by TEM in situ micropillar compression deformation analysis using the Fe alloy single crystal (Zhang et al. 2012). Figure 8.14a, b shows snapshots extracted from movies of deformation during the tests. Figure 8.14c shows the load–displacement data recorded in synchronism with the observation, as well as the relation between the dislocation behavior and mechanical response. The compression axis of the pillar is parallel to the <110> axis, and the directions in which the two <111> directions that can be potential. Burgers vectors are indicated by arrows in Fig. 8.14a, b. The dislocation component is roughly judged from the relation between a line vector of the dislocation and the direction of the Burgers vector <111>. The screw component dominates when the dislocation line is parallel to the compression axis of the pillar, while the edge dislocation dominates the perpendicular dislocation line. In the dislocation shown in Fig. 8.14a, the edge component is judged to be dominant from the direction of the dislocation line. The two images (a1) and (a2) shown in Fig. 8.14a are taken at 1/30 s intervals. During the interval, the edge dislocation moves instantaneously from the left end to the right end in the figure, indicating a very high movement speed. The moment this phenomenon is observed, the corresponding mechanical behavior is indicated by arrow 1 on the load–displacement curve, which is the first half of the relatively lower stress deformation. On the other hand, in the dislocation shown in (b), as the dislocation line is almost parallel to the <111> axis, it can be judged that the screw component is dominant. Comparing (b2) and (b1) recorded at intervals of approximately 3 s, to examine the moving speed,

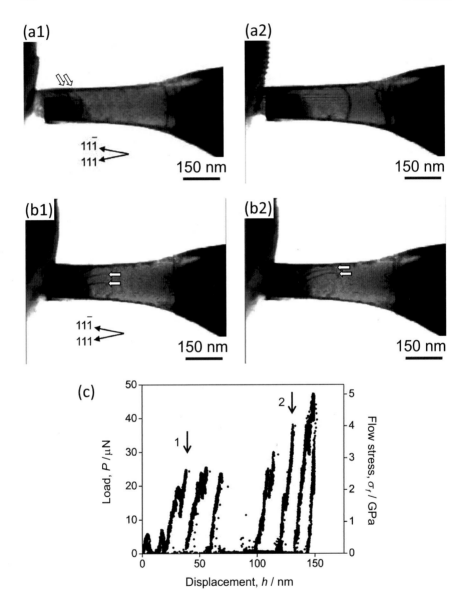

Fig. 8.14 **a** and **b** are snapshots extracted from movies of deformation during the tests. Figure 8.10c shows the load–displacement data recorded in synchronism with the observation (Zhang et al. 2012). Reprinted with permission from [L. Zhang, T. Ohmura, K. Sekido, T. Hara, K. Nakajima and K. Tsuzaki: Scripta Mater., 67 (2012), 388–391.] Copyright (2012) by Elsevier

while the time intervals are nearly 10 times as long as those of (a1) and (a2), the travel distance during the time interval is small and the moving speed is much lower than the edge dislocation. The mechanical behavior corresponding to the moment with screw dislocation dominancy is the position on the load–displacement curve, indicated by arrow 2 in (c), and the flow stress is higher than that in the case of edge dislocation in (a). These results indicate that the mobility of screw dislocations in bcc metals is extremely lower than that of edge dislocations, and the flow stress is thereby increased.

In addition, the TEM in situ deformation analysis of IF steel successfully captures the dynamic relation between dislocation density and flow stress under the condition dominated by screw dislocations (Zhang et al. 2014). The sample exhibits a blade-like shape with a height and width of approximately 600 nm and a thickness of approximately 80 nm. Figure 8.15a shows the true stress–true strain curves measured simultaneously with TEM observation with two curves to show reproducibility. Figure 8.15b–f shows snapshots taken from a movie recorded by TEM in situ deformation. The compression axis is horizontal, and a diamond flat-end punch approaches the specimen from the left side of the figure. As shown in Fig. 8.15b, the dislocation density before the yield stress is extremely low. The yield stress obtained from (a) exceeds 1.0 GPa, which is orders of magnitude higher than the bulk yield stress. This is consistent with the size-dependent behavior described in Sect. 8.4.

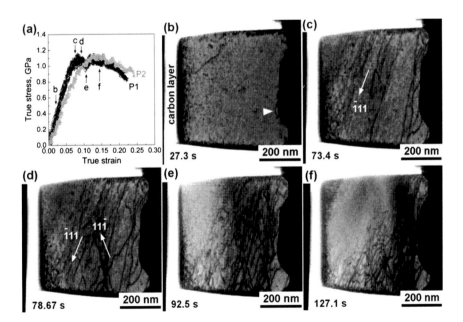

Fig. 8.15 a True stress—true strain curves measured simultaneously with TEM observation. **b–f** are snapshots taken from a movie recorded by TEM in situ deformation (Zhang et al. 2014). Reprinted with permission from [L. Zhang, N. Sekido and T. Ohmura: Mater. Sci. Eng., **A611** (2014), 188–193.] Copyright (2014) by Elsevier

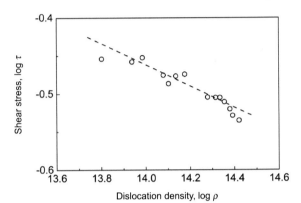

Fig. 8.16 Results of fitting by Eq. (8.16) for changes in flow stress obtained from the stress–strain curve and dislocation density obtained from TEM images (Zhang et al. 2014). Reprinted with permission from [L. Zhang, N. Sekido and T. Ohmura: Mater. Sci. Eng., A611 (2014), 188–193.] Copyright (2014) by Elsevier

After yielding, the stress tends to decrease gradually. At the same time, the dislocation density is observed to increase gradually. The relation between flow stress and dislocation density is discussed below. In this case, unlike the case of Fig. 8.2, the increase in dislocation density is slow, and thus the number of dislocations for a short time can be regarded as constant. The deformation is under the dislocation mobility, not under the growth of dislocation density, and the relation between the macroscopic plastic strain rate and dislocation density is discussed using Eq. (8.12). As the relation between the dislocation mobility and the applied stress is given by Eq. (8.14), under the condition that the moving speed of dislocations, which give plastic deformation, is the same, the relation between the shear stress and dislocation density is arranged as follows:

$$\log \tau \propto A - \frac{1}{m} \log \rho, \tag{8.16}$$

where A is a constant. Figure 8.16 shows the results of fitting by Eq. (8.16) for the changes in flow stress obtained from the stress–strain curve and dislocation density obtained from TEM images. From this plot, the exponent m value is determined to be approximately 7. According to the literature (Stein and Low 1960), the m value obtained from experimentally determining the relation between shear stress and the mobility in an edge dislocation dominancy in Fe-Si alloy is approximately 40, which is 6 times larger than the result shown in Fig. 8.16. This is attributed to the fact that the mobility of screw dislocations in bcc is lower than that of edge dislocations, which is consistent with our understanding of the mobility of screw dislocations. Additionally, the values of the stress exponent m are discussed as follows. In Eq. (8.15), under the condition that the dislocation density is constant, the relation between the strain rate and stress can be expressed as follows:

$$\frac{\partial \ln \dot{\gamma}}{\partial \tau} = \frac{m}{\tau}. \tag{8.17}$$

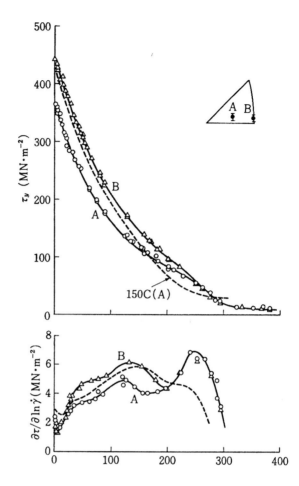

Fig. 8.17 Results of the yield stress of pure iron single crystal and its strain rate dependence (Aono et al. 1981). Reprinted with permission from [Y. Aono, E. Kuramoto and K. Kitajima: Rep. Res. Inst. Appl. Mechanics, Kyu-shu Univ. XXIX, (1981), 127.] Copyright (1981) by Research Institute for Applied Mechanics, Kyushu University

On the other hand, Fig. 8.17 shows the results of the yield stress of pure iron single crystal and its strain rate dependence (Aono et al. 1981). The yield stress at room temperature and its strain rate dependence can be read as 20 and 3 MPa, respectively. Substituting them into Eq. (8.17), the m value is calculated to be approximately 7. Note that this value agrees well with that obtained from the plot shown in Fig. 8.16. These results are the first real-time demonstration of the Johnston–Gilman model, which expresses the relation between dislocation density and flow stress, and it can be said that the elementary process of the relation between dislocation behavior and mechanical response is approached by the new observation and analysis technology.

8.4.3 Plasticity Induced by Phase Transformation

Indentation-induced phase transformation is another important behavior during nanoindentation, as reported for various materials (Ahn et al. 2010; Crone et al. 2007; Frick et al. 2006). Phase transformation is detected by SPM imaging of the sample surface after nanoindentation in most cases, and mechanical behavior analysis for the $P–h$ relation is an important approach for investigating this behavior. As a new analytical approach, a transition in the plastic deformation mechanisms is analyzed from the $P/h–h$ plots (Sekido et al. 2011b). The theoretical load with a conical or pyramidal indenter in an elastoplastic deformation is given by the following equation:

$$P_t = ah^2, \tag{8.18}$$

where a is the material constant that depends on the elastic and the plastic properties of a material. Thus, $P/h–h$ plots should show a constant slope corresponding to a when the deformation mode is kept the same during the deformation. Although the actually measured P_m includes the influences of the tip truncation and stiffness of the load frame, expressed by

$$P_m = ah^2 + a_2h, \tag{8.19}$$

where $a2$ is the constant that is corresponding to the shape of the indenter tip and the stiffness of the load frame (Ohmura et al. 2002). The coefficient $a2$ turns into the y-intercept by the transformation to the $P/h–h$ plots to separate from the parameter a that corresponds to the intrinsic behavior of materials. However, parameter a changes when different deformation modes operate during a plastic deformation. Then, Eq. (8.18) is expressed in the following two ways:

$$P_t = a_e h_e^2 \tag{8.20}$$

and

$$P_t = a_p h_p^2, \tag{8.21}$$

where a_e and a_p are constants for elastic and plastic deformation, and h_e and h_p are the elastic and plastic penetration depths, respectively. a_p is proportional to hardness H given by $H = P/A$, where P is the applied load and A is the projected area of contact, which is proportional to h_p^2. As h can be expressed through a simple summation of h_e and h_p (Oliver and Pharr 1992), the relation among a, a_e, and a_p is given by

$$a^{-1/2} = a_e^{-1/2} + a_p^{-1/2}. \tag{8.22}$$

Assuming that a_e, which is correlated with Young's modulus, is not associated with plastic strain, only a_p is the controlling factor for a in Eq. (8.18). Hence, a slope change on a P/h–h plot corresponding to a in Eq. (8.19) denotes a change in a_p, which is affected by the plastic deformation mode and indicates that a load for the transition of deformation modes can be visualized by the slope change on the P/h–h curve.

Figure 8.18a–c represents the typical P–h curves obtained from the nanoindentation tests under peak loads of 1, 4, and 8 mN. A sudden displacement burst, which is called a pop-in, is observed on the P–h curves, shown as insets in the upper left corner of the figure. The critical load at which the first pop-in occurs is indicated as P_c and the corresponding excursion depth is denoted as Δh. The broken lines in the three figures correspond to the calculated curves from Eq. (8.1), which are obtained by substituting $E^* = 200$ GPa and $R_i = 230$ nm. The experimental data below P_c agree with the calculated curves, suggesting a purely elastic deformation below P_c. Figure 8.18d–f exhibits the P/h–h plots calculated from Fig. 8.18a–c, respectively. The slopes of the P/h–h plots are approximately 0.06 μN/nm^2 at an early stage of plastic deformation, following the first pop-ins that occur below 1 mN, as shown in Fig. 8.18d–f. The slopes of approximately 0.10 μN/nm^2 appear through a further loading of up to 4 and 8 mN, as shown in Fig. 8.18e, f. This slope change on the

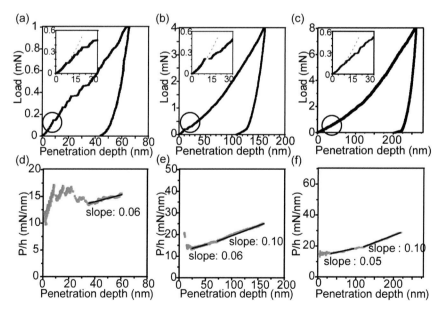

Fig. 8.18 P–h curves obtained by nanoindentation with peak loads of **a** 1 mN, **b** 4 mN, and **c** 8 mN. The P–h curves (**a**), (**b**), and (**c**) after the first pop-ins are converted into P/h versus h curves of **d** 1 mN, **e** 4 mN, and **f** 8 mN, respectively (Sekido et al. 2011b). Reprinted with permission from [K. Sekido, T. Ohmura, T. Sawaguchi, M. Koyama, H.W. Park and K. Tsuzaki: Scripta Mater., 65 (2011), 942–945.] Copyright (2011) by Elsevier

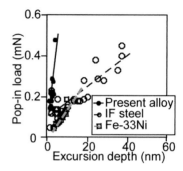

Fig. 8.19 Relationship between P_c and Δh for the first pop-ins (Sekido et al. 2011b). Reprinted with permission from [Ohmura, T. Sawaguchi, M. Koyama, H.W. Park and K. Tsuzaki: Scripta Mater., 65 (2011), 942–945.] Copyright (2011) by Elsevier

P/h–h curves is speculated to be the consequence of the change in the predominant deformation mode.

The SPM observations indicate that the stress-induced ε-martensitic transformation is the deformation mode that predominantly operates at an early stage of plastic deformation during nanoindentation (Sekido et al. 2011b). Figure 8.19 shows the relation between the critical load P_c and the corresponding excursion depth Δh at the first pop-in. The results for an Fe-33Ni steel (FCC) and a Ti-added IF steel (BCC) are also shown for comparison. The figure shows that based on equivalent P_c values, the values of Δh in the present alloy are significantly smaller than those in the IF and Fe–Ni steels. P_c and Δh have been shown to be closely related to their Young's moduli (Ohmura and Tsuzaki 2007b); however, Young's modulus of the present alloy is comparable to that of the IF and Fe–Ni steels, suggesting that the smaller values of Δh cannot be attributed to the elastic property. Note that the present alloy shows the stress-induced ε-martensitic transformation at a lower load (Otsuka et al. 1990), while the slip deformation has been identified as the predominant deformation mode in IF and Fe–Ni steels, at least on a bulk scale. Therefore, a smaller Δh is derived from the deformation mode operating in the present alloy, i.e., the stress-induced ε-martensitic transformation. Using the geometrically necessary (GN) dislocation loop model proposed by Shibutani et al. (Shibutani et al. 2007), the excursion depth Δh is given as

$$\Delta h = nb, \tag{8.23}$$

where n is the number of GN dislocations. Equation (8.23) shows that a smaller Δh corresponds to fewer dislocations formed during the pop-in event. ε-martensite is known to form by the motion of partial dislocations on alternative {111} planes. As the dislocation multiplication in FCC metals requires the shrinkage of extended dislocations, no dislocation multiplication occurs during the stress-induced ε-martensitic transformation. Thus, the smaller Δh in the present alloy also implies the occurrence of ε-martensitic transformation.

In a practical design of high-performance steel by phase transformation, the transformation-induced plasticity (TRIP) effect is a major factor for maintaining a good balance between strength and elongation (Cooman 2004; Aydin et al. 2013; Cao et al. 2011; Zaefferer et al. 2004; Jacques 2004; Zackay et al. 1967). The stability of austenite phase is affected by many factors, including chemical composition, grain size, grain geometrical shape, crystallographic orientation, and the phase surrounding the retained austenite, and further understanding of individual factors is required. To address this issue, nanoindentation techniques are applied to investigate the mechanical stability of individual retained austenite grains in TRIP steels, especially for the boundary effect.

Nanoindentation tests are performed for the austenite grains with different size in high-carbon quenched-tempered steel (Man et al. 2019). Figure 8.20 presents the phase maps and SPM images of the three austenite grains, which is highlighted by dashed lines. The corresponding $P/h–h$ plots for the three grains are presented in Fig. 8.21. Interestingly, all $P/h–h$ plots show double stages on the loading curve during the plastic deformation, i.e., slope a with a low value for in stage I turns into a higher value in stage II. In addition, the value of a in stage I is higher for the small grains. Furthermore, the transition load P_t, which is defined as the change in the slope of the $P/h–h$ plot, is found to increase with decreasing the austenite grain size. In fact, the slope a in stage I is lower and the P_t value is relatively clear for a large austenite grain. In contrast, a is larger in stage I, and thus P_t is difficult to determine for a small austenite grain. For getting a reliable conclusion, the nanoindentation tests were repeated for 76 austenite grains with different sizes. Thus, Fig. 8.22a, b shows plots of slope a in stage I and the P_t value, respectively, as a function of austenite grain size D. As indicated, when the austenite grain size increases, both the slope a in stage I and the P_t value decrease. Although the data show some scattering due to the irregular shapes of the austenite grains, it is concluded that larger grain size has lower resistance to plastic deformation. A change in the slope of the $P/h–h$ plot is believed to be the consequence of a change in the deformation mode for the plastic deformation in the retained austenite, which is a transition from the stress/strain-induced martensitic transformation of the metastable retained austenite into the dislocation glide motion in the transformed martensite. It is also found that both the slope a in stage I and the value of P_t increase upon decreasing the austenite grain size. This result suggests that the mechanical stability of the retained austenite increases with decreasing grain size, that is, the resistance against stress-induced martensitic transformation increases in smaller grained austenite. This can be partly attributed to a constraint effect by the surrounding tempered martensite phase. As the tempered martensite phase is harder than the retained austenite phase, the austenite to martensite transformation with volume expansion should be inhibited by the tempered martensite phase. This effect is more significant in a region close to the interface because the volume expression by the γ to α' transformation is subjected to greater compressive stresses and space limitations from the martensite–austenite interface in smaller austenite grains.

The boundary effect on the mechanical stability of metastable austenite (γ) in the Fe–Ni steels is characterized using a combination of nanoindentation and transmission electron microscopy (Man et al. 2020). Figure 8.23a, c shows phase maps drawn after the nanoindentation tests, where red and green indicate γ and α', respectively. Figure 8.23b, d shows the corresponding Image Quality maps. Nanoindentation tests

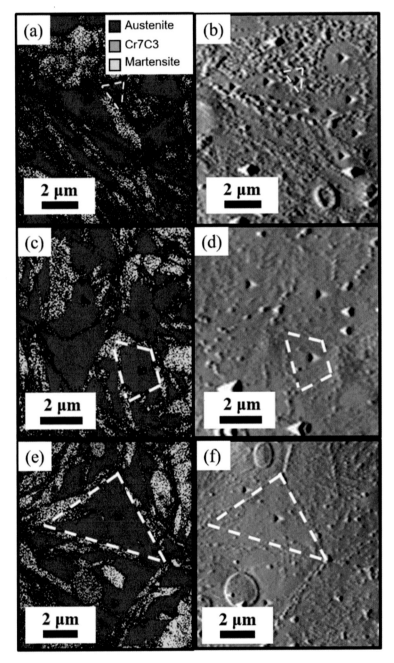

Fig. 8.20 **a** Phase map and **b** SPM image showing the indentation mark position in the retained austenite grain with a smaller grain size of 1.55 μm indicated by dashed triangle. **c, d** The middle size of 2.76 μm indicated by dashed quadrangle, and **e, f** a larger size of 5.38 μm indicated by dashed triangle (Man et al. 2019). Reprinted with permission from [T. Man, T. Ohmura and Y. Tomota: ISIJ Int., 59 (2019), 559–566.] Copyright (2019) by The Iron and Steel Institute of Japan

Fig. 8.21 *P/h* versus *h* plots corresponding to the three grains with sizes of **a** 1.55 μm, **b** 2.76 μm, **c** 5.38 μm shown in Fig. 8.6. All plots exhibit the two stages of I (blue) and II (green) (Man et al. 2019). Reprinted with permission from [T. Man, T. Ohmura and Y. Tomota: ISIJ Int., 59 (2019), 559–566.] Copyright (2019) by The Iron and Steel Institute of Japan

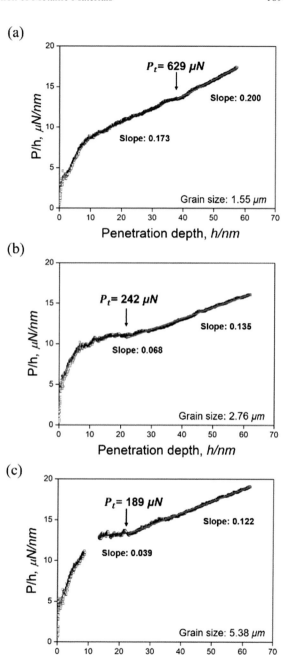

Fig. 8.22 a Slope *a* in stage
I and **b** the transition load P_t
plotted as a function of the
grain size *D* of retained
austenite phase (Man et al.
2019). Reprinted with
permission from [T. Man, T.
Ohmura and Y. Tomota: ISIJ
Int., 59 (2019), 559–566.]
Copyright (2019) by The
Iron and Steel Institute of
Japan

(a)

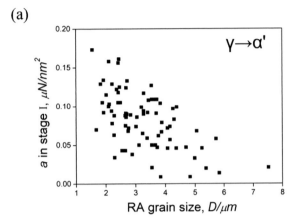

(b)

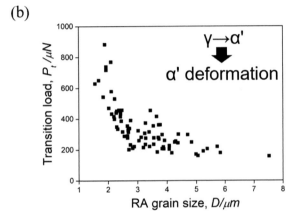

with a peak load of 2000 μN are conducted within the γ grain interior and in the vicinity of the γ/γ grain boundary, as well as the γ/α' interface. The indentation marks at the γ grain interior, in the vicinity of the γ/γ grain boundary, as well as the γ/α' interface, are outlined by pink, orange, and blue circles and arrows, respectively, as shown in Fig. 8.23b, d. Figure 8.24 shows a plot of the average values of slope *a* on $P/h - h$ plot in stage I and P_t for metastable γ in Fe-27Ni and the stable γ in Fe-30Ni. The plot includes the results for the γ grain interior, γ/γ grain boundary, and γ/α' interface, which are indicated by rectangles, circles, and rhombi, respectively. In each case, the error bars are calculated on the basis of SDs for the total data. The results show a tendency for the average values of slope *a* in stage I and of P_t for metastable γ to become lower for the γ/α' interface, γ/γ grain boundary, and γ grain interior, in turn. Furthermore, the average slope of the *a* value for the γ grain interior is lower in stage I of metastable γ in Fe-27Ni (0.013 μN/nm²) than in the plastic deformation stage of stable γ in Fe-30Ni (0.025 μN/nm²). In addition, the difference in the average slope values between the γ/γ grain boundary and γ grain interior is higher for metastable γ in Fe-27Ni (0.012 μN/nm²) than that for stable γ in Fe-30Ni

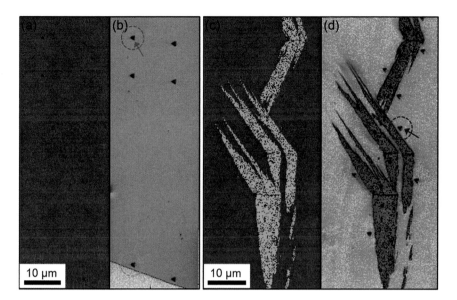

Fig. 8.23 a, c Phase maps, where red indicates austenite (γ) and green indicates martensite (α'), and **b, d** IQ maps taken after nanoindentation tests with a peak load of 2000 μN. The indentation marks at the γ grain interior and in the vicinity of the γ/γ grain boundary in Fe-27Ni-H, and in the vicinity of the γ/α' interface in Fe-27Ni-S, are outlined by the pink, orange, and blue circles and arrows in (**b**) and (**d**), respectively (Man et al. 2020). Reprinted with permission from [T. Man, T. Ohmura and Y. Tomota: Mater. Today Comm., 23 (2020), 100896] Copyright (2020) by Elsevier

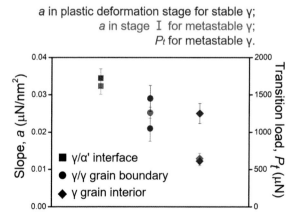

Fig. 8.24 Plot of average values of slope a in stage I and P_t for metastable γ in Fe-27Ni-H and Fe-27Ni-S, and slope a in plastic deformation stage for stable γ in Fe-30Ni, where γ grain interior, γ/γ grain boundary, and γ/α' interface are indicated by rectangles, circles, and rhombi, respectively. Error bars are calculated based on standard deviation for total data in each case (Man et al. 2020). Reprinted with permission from [T. Man, T. Ohmura and Y. Tomota: Mater. Today Comm., 23 (2020), 100896] Copyright (2020) by Elsevier

(0.004 μN/nm^2). The average values of slope a in stage I and P_t for the γ/γ grain boundary and γ/α' interface are higher than those for the γ grain interior, indicating that the resistance of the boundaries to martensitic transformation is higher than that in the γ grain interior. Furthermore, the average values of slope a in stage I and P_t for the γ/α' interface are higher than those for the γ/γ grain boundary, suggesting that the constraint effect of the γ/α' interface is higher than that of the γ/γ grain boundary, as the hardness of α' is higher than that of γ.

8.5 Summary

This chapter introduces nanoindentation and TEM in situ deformation analysis, which are a method for directly analyzing the mechanical behavior on nanoscale and capturing the relation with dislocation motion in real time. The new analysis method enables us to grasp the behavior on the nanoscale in detail, and new knowledge is obtained on the function of grain boundaries as dislocation sources, the deformation behavior related to dislocation motion other than viscous motion, and the relation between the character of dislocation and mechanical response. To clarify the mechanism of plastic phenomena in more detail, it is necessary to reveal the effect of temperature, which is an important external state variable, as well as stress, and it is desired to improve the measuring technique at low temperatures below room temperature, or conversely, at high temperatures. To overcome the 10^6 order gap mentioned in the Introduction, it is necessary to construct a material behavior model based on the new knowledge. This is a barrier that could not be completely overcome even by the conventional dislocation theory, but we would like to continue challenging it further through advanced efforts, such as new experimental analysis methods.

A new concept of "*Plaston*" was proposed for further understanding the mechanical behavior and controlling the performance of structural materials (Tsuji et al. 2020). *Plaston* presumably include a singularity with a stress intensity and an excited state with elastic strain energy leading to a mechanically instability in a local atomistic arrangement. It is expected that the unstable state instantaneously transfers to metastable states in various lattice defects including dislocations, twin defects, cracks, phase boundaries and so on. As the transferred defects subsequently dominate the macroscopic mechanical properties of materials, it is important to control a transition path to choose an appropriate defect structure. That is a novel guide principle to get high-performance materials. The mechanical singularity presumably occurs at local regions such as crack tip, grain boundary, triple junction etc., which can hardly be captured by conventional experimental approach. For the issuers, the nanoindentation has a strong potential by inducing an excited state intentionally at any positions in a material to see how materials behave under the unstable state. In particular, the pop-in event, which is described in this chapter, is an interesting phenomenon under a remarkably high-stress state close to the theoretical strength level. Therefore, nanomechanical characterization has a great potential to reveal an elementally step of various deformation modes and develop the concept of *Plaston*.

Acknowledgements This work was supported by the Elements Strategy Initiative for Structural Materials (ESISM) of MEXT (Grant number JPMXP0112101000).

References

Ahn TH, Oh CS, Kim DH, Oh KH, Bei H, George EP, Han HN (2010) Scripta Mater 63:540–543
Aono Y, Kuramoto E, Kitajima K (1981) Rep Res Inst Appl Mech Kyu-shu Univ XXIX:127
Aydin H, Essadiqi E, Jung I, Yue S (2013) Mater Sci Eng A 564:501
Bolshakov A, Pharr GM (1998) J Mater Res 13:1049
Brenner SS (1956) J Appl Phys 27:1984
Britton TB, Randman D, Wilkinson AJ (2009) J Mater Res 24:607–615
Cao WQ, Wang C, Shi J, Wang MQ, Hui WJ, Dong H (2011) Mater Sci Eng A 528:6661
Carpeno DF, Ohmura T, Zhang L, Leveneur J, Dickinson M, Seal C, Kennedy J, Hyland M (2015) Mater Sci Eng A639:54–64
Carrington WE, McLean D (1965) Acta Matell 13:493–499
Chung T-F, Yang Y-L, Huang B-M, Shi Z, Lin J, Ohmura T, Yang J-R (2018) Acta Mater 149:377–387
Cottrell AH, Bilby BA (1949) Proc Phys Soc Sect A 62:49–62
Crone WC, Brock H, Creuziger A (2007) Exp Mech 47:133–142
De Cooman BC (2004) Curr Opin Solid State Mater Sci 8:285
Doerner MF, Nix WD (1986) J Mater Res 1:601
Edagawa K, Suzuki T, Takeuchi S (1997) Phy Rev B 55:6180
Frick CP, Lang TW, Spark K, Gall K (2006) Acta Mater 54:2223–2234
Gerberich WW, Nelson JC, Lilleodden ET, Anderson P, Wyrobek JT (1996) Acta Mater 44:3585–3598
Gouldstone A, Koh H-J, Zeng K-Y, Giannakopoulos AE, Suresh S (2000) Acta Mater 48:2277–2295
Hall EO (1951) Proc R Soc B64:747–753
Hauser JJ, Chalmers B (1961) Acta Matell. 9:802–818
Hirsch PB (1968) Suppl Trans JIM 9:XXX
Hsieh Y-C, Zhang L, Chung T-F, Tsai Y-T, Yang J-R, Ohmura T, Suzuki T (2016) Scripta Mater 125:44–48
Itokazu M, Murakami Y (1993) Trans Jpn Soc Mech Eng A 59:560
Jacques PJ (2004) Curr Opin Solid State Mater Sci 8:259
Johnson KL (1985) Contact mechanics. Cambridge University Press, Cambridge, UK, pp 84–106
Johnston WG (1962) J Appl Phys 33:2716
Johnston WG, Gillman JJ (1959) J Appl Phys 30:129
Kim Y-J, Yoo B-G, Choi I-C, Seok M-Y, Kim J-Y, Ohmura T, Jang J-I (2012) Mater Lett 75:107–110
Kurzydlowski KJ, Varin RA, Zielinski W (1984) Acta Metall 32:71–78
Lee TC, Robertson IM, Birnbaum HK (1990) Met Trans 21A:2437–2447
Leipner HS, Lorenz D, Zeckzer A, Lei H, Grau P (2001) Physica B 308–310:446–449
Li JCM (1963) Trans AIME 227:239–247
Lim YY, Chaudhri M (1999) Phil Mag A 79:2979
Lorenz D, Zeckzer A, Hilpert U, Grau P, Johansen H, Leipner HS (2003) Phys Rev B 67:172101
Loubet JL, Georges JM, Marchesini JM, Meille G (1984) J Tribl 106:43
Man T, Ohmura T, Tomota Y (2019) ISIJ Int 59:559–566
Man T, Ohmura T, Tomota Y (2020) Mater Today Comm 23:100896
Masuda H, Morita K, Ohmura T (2020) Acta Mater 184:59–68
McElhaney KW, Vlassak JJ, Nix WD (1998) J Mater Res 13:1300
Minor AM, Asif SAS, Shan Z, Stach EA, Cyrankowski E, Wyrobek TJ, Warren OL (2006) Nat Mater 5:697–702

Nakano K, Ohmura T (2020) J Iron Steel Inst Japan 106:82–91
Newey D, Willkins MA, Pollock HM (1982) J Phys E Sci Instrum 15:119
Nishibori M, Kinoshita K (1978) Thin Solid Films 48:325
Nishibori M, Kinoshita K (1997) Japan. J Appl Phys 11:758
Nix WD, Gao H (1998) J Mech Phys Solids 46:411
Ohmura T, Tsuzaki K (2007a) Materia Japan 46:251
Ohmura T, Tsuzaki K (2007b) J Mater Sci 42:1728–1732
Ohmura T, Tsuzaki K (2008) J Phys D 41:1–6
Ohmura T, Matsuoka S, Tanaka K, Yoshida T (2001) Thin Solid Films 385:198–204
Ohmura T, Tsuzaki K, Matsuoka S (2002) Phil Mag A 82:1903–1910
Ohmura T, Tsuzaki K, Tsuji N, Kamikawa N (2004a) J Mater Res 19:347–350
Ohmura T, Minor A, Stach E, Morris JW Jr (2004b) J Mater Res 19:3626–3632
Ohmura T, Tsuzaki K, Yin F (2005) Mater Trans 46:2026–2029
Ohmura T, Zhang L, Sekido K, Tsuzaki K (2012) J Mater Res 27:1742–1749
Oliver WC, Pharr GM (1992) J Mater Res 7:1564
Onose K, Kuramoto S, Suzuki T, Chang Y-L, Nakagawa E, Ohmura T (2019) J Jpn Inst Light Metals
    69:273–280
Otsuka H, Yamada H, Maruyama T, Tanahashi H, Matsuda S, Murakami M (1990) ISIJ Int 30:674
Petch NJ (1953) J Iron Steel Inst 174:25–28
Sato Y, Shinzato S, Ohmura T, Ogata S (2019) Int J Plast (In Press)
Schuh CA, Mason JK, Lund AC (2005) Nat Mater 4:617
Sekido K, Ohmura T, Zhang L, Hara T, Tsuzaki K (2011a) Mater Sci Eng A 530:396–401
Sekido K, Ohmura T, Sawaguchi T, Koyama M, Park HW, Tsuzaki K (2011b) Scripta Mater 65:942–
    945
Sekido K, Ohmura T, Hara T, Tsuzaki K (2012) Mater Trans 53:907–912
Shen Z, Wagoner RH, Clark WAT (1988) Acta Metall 36:3231–3242
Shibutani Y, Koyama A (2004) J Mater Res 19:183–188
Shibutani Y, Tsuru T, Koyama A (2007) Acta Mater 55:1813
Stein DF, Low JR (1960) J Appl Phys 31:362
Suzuki H (1968) Dislocation dynamics, Rosenfield AR, Hahn GT, Bement AL, Jaffee RI (eds).
    McGraw Hill, New York:679–700
Suzuki T,. TakeuchiT (1989) Lattice defects in ceramics, Takeuchi S, Suzuki T (eds). Publication
    office of Japanese journal of applied physics. Tokyo, p 9
Takeuchi S (1981) Intermetallic potentials and crystalline defects, Lee JK (ed). The metallurgical
    society of AIME, pp 201–221
Takeuchi S, Suzuki T (1989) Strength of metals and alloys, Kettunen PO, Lepisto TK, Lehtonen
    ME (eds). Pergamon Press, Oxford, p 161
Tsui TY, Oliver WC, Pharr GM (1996) J Mater Res 11:752
Tsuji N, Ogata S, Inui H, Tanaka I, Kishida K, Gao S, Mao W, Bai Y, Zheng R, Du J-P (2020)
    Scripta Mater 181:35
Uchic MD, Dimiduk DM, Florando JN, Nix WD (2004) Science 305:986
Ushioda K, Takata K, Takahashi J, Kinoshita K, Sawada H (2019) J Jpn Inst Metals 83:353 (in
    Japanese)
Vitek V (1974) Cryst Lattice Defects 5:1
Zackay VF, Parker ER, Fahr D, Busch R (1967) Trans ASM 60:252
Zaefferer S, Ohlert J, Bleck W (2004) Acta Mater 52:2765
Zbib AA, Bahr DF (2007) Metall. Mater Trans 38A:2249–2255
Zhang L, Ohmura T, Emura S, Sekido N, Yin F, Min X, Tsuzaki K (2009) J Mater Res 24:2917–2923
Zhang L, Ohmura T, Sekido K, Nakajima K, Hara T, Tsuzaki K (2011) Scripta Mater 64:919–922
Zhang L, Ohmura T, Sekido K, Hara T, Nakajima K, Tsuzaki K (2012) Scripta Mater 67:388–391
Zhang L, Ohmura T (2014) Phys Rev Lett 112:145504
Zhang L, Sekido N, Ohmura T (2014) Mater Sci Eng A611:188–193

Chapter 9
Synchrotron X-ray Study on Plaston in Metals

Hiroki Adachi

Methods that can be used to reinforce metal include solid solution strengthening (Fleischer 1963), precipitation strengthening (Gerold and Harberkorn 1966), dislocation hardening (Bailey and Hirsch 1960), and grain refining (Petch 1953; Hall 1951). In particular, in grain refining, as expressed in the Hall–Petch equation, the strength of the material increases linearly as the crystal grain becomes finer. As the process does not inevitably require the addition of many elements, it is suitable for use in recycling, lessens the load on the environment, and consequently has recently attracted attention as a method for reinforcing structural metal materials. Another reason for the increased focus on grain refining is that the development of a severest plastic deformation method allows a relatively easy preparation of submicron grain metal crystals (Valiev et al. 2000; Tsuji et al. 2002; Huang et al. 2003; Horita and Langdon 2005; Ferrasse 1997).

The ultra-fine grained (UFG) materials thus obtained exhibit significantly high strength as well as unique mechanical characteristics, such as the extra-hardening phenomenon (Kamikawa et al. 2003, 2009), the hardening-by-annealing phenomenon (Huang et al. 2006), and the yield-point-drop phenomenon (Kamikawa et al. 2003, 2009), in AI alloys. In the extra-hardening phenomenon, when the crystal grain size falls below a few micrometers, the strength of the UFG material increases beyond that represented by a line extended from the slope of the Hall–Petch plot. Although with coarse-grained (CG) materials, the hardening-by-annealing phenomenon results in recovery and recrystallization, and since there is decreased dislocation density, which results in decreased strength and increased ductility, with UFG materials, this phenomenon causes an entirely opposite change of increasing the strength of the material and decreasing its ductility. Moreover, while

H. Adachi (✉)
Department of Materials and Synchrotron Radiation Engineering, University of Hyogo, Himeji, Japan
e-mail: adachi@eng.u-hyogo.ac.jp

the yield-point-drop phenomenon is observed in steel materials with a BCC structure, normally, because Al alloys with a face-centered cubic (FCC) structure show a continuous decrease, they show no yield-point-drop phenomenon. However, yield point drop has been reported in UFG Al alloys. Otherwise as well, for super-fine-grain materials, it has been reported that the strain rate dependence of yield strength becomes very large.

As aforementioned, UFG materials exhibit unique mechanical characteristics, believed to be a result of their mechanism of plastic deformation being different from that of CG materials. One reason for this difference is thought to be their significantly high grain densities compared to those of CG materials. Plastic deformation in CG materials occurs due to dislocations, which gradually increase inside grains. However, in UFG materials, deformation also occurs due to the formation of nanotwin crystals generated from the grain boundary and stacking faults. Another reason is thought to be the difference in the behavior of dislocations inside the grains from CG materials. We referred the components of deformation as "*Plaston*," and this study aims to improve the strength and ductility of metallic materials by understanding plastons.

The first approach is to understand the extent to which the unique mechanical characteristics of UFG materials can be explained by dislocation motion and to which they cannot. That is, there is a need to understand whether the characteristics must be explained using conceptions of plastons other than dislocation. It has been indicated that nanotwins and stacking faults can occur at grain boundaries in UFG materials (Lu et al. 2005; Shen et al. 2005; Chen et al. 2003), so dislocation can both occur and disappear at the grain boundaries (Mompiou et al. 2012). For this reason, it is necessary to conduct research while considering the possibility that the dislocation substructure might differ during deformation and after unloading, and thus it is desirable to conduct in situ measurements. To date, there have been examples described in the literature where the behavior of dislocations during deformation was observed using a transmission electron microscope (TEM). However, the films for TEM observations are extremely thin and the surface effect on the dislocation motion cannot be neglected. Therefore, it is desirable to use bulk materials for such measurements. For this reason, in situ X-ray diffraction (XRD) measurements were conducted during the deformation of materials at SPring-8, the largest synchrotron radiation facility, and the effect of the crystal grain size on the dislocation behavior was studied (Adachi et al. 2015, 2016; Miyajima et al. 2016; Nakayama et al. 2016), [39].

The synchrotron radiation at SPring-8 exhibits high flux, which enables the measurement of the diffraction intensity with a high signal-to-noise (S/N) ratio within a short time. Furthermore, a detector with a large area enables multiple diffraction peaks to be collected simultaneously, making it possible to carry out in situ measurements with high temporal resolution and record diffraction peaks over a large diffraction angle range within ~0.5 s. Additionally, synchrotron radiation exhibits high directivity, which reduces the effect of the instrumental function on the XRD measurement results and represents an advantage of synchrotron radiation. The Williamson–Hall method is used for calculating the dislocation density from the obtained XRD diffraction profile, expressed as

$$\frac{\Delta2\theta_{hkl}cos\theta}{\lambda} = \frac{0.9}{D} + 2\epsilon\frac{sin\theta_{hkl}}{\lambda},$$

where θ is the diffraction peak angle, $\Delta2\theta$ is the full width at half maximum, and D is the crystallite size, which can be obtained from the reciprocal of the intercept of the Williamson–Hall plot, where θ and $\Delta2\theta$ values of multiple diffraction peaks from the XRD measurements are used to construct a Williamson–Hall plot, with $2sin\theta/\lambda$ on the horizontal axis and $\Delta2\theta cos\theta/\lambda$ on the vertical axis. The inhomogeneous strain, ϵ, in the crystallite can be obtained from the slope of the plot. Assuming that the inhomogeneous strain is a result of dislocation, the dislocation density can be calculated using either of the following equations, (9.1) or (9.2). Here, the coefficient in Eq. (9.1) has a value of 16.1 for the FCC structure and that of 14.4 for the BCC structure. Since the goal here is to compare CG and UFG materials, Eq. (9.1) is chosen because it is difficult to accurately obtain the crystallite size, D, for CG materials, which is a required parameter in Eq. (9.2) (Williamson and Hall 1953; Williamson and Smallman 1956, 1954).

$$\rho = 16.1\left(\frac{\epsilon}{b}\right)^2 \tag{9.1}$$

$$\rho = \frac{2\sqrt{3}\epsilon}{Db}. \tag{9.2}$$

Figure 9.1 shows the change in the dislocation density during tensile deformation in a 2 N-aluminum alloy with a crystal grain size of 0.5 μm, formed using the ARB method. As the deformation progresses, the dislocation density changes through four regions. In the first region (region I), the dislocation density does not increase. As the stress increases linearly with macrostrain, it is said to be an elastic deformation region. In the second region (region II), the dislocation density increases rapidly, showing the start of plastic deformation, where stress at this time is denoted by σI. For this reason, the stress σ_I at the boundary between regions I and II can be recognized as the yield stress when the dislocation starts to increase from the dislocation source. The increase in dislocations in region II is almost linear with that in the macrostrain, but the system transitions to the third region (region III) when the dislocation density reaches a certain value (ρ_{II}). The increase in dislocations in region III is slower than that in region II. However, it is not that the rapid increase in the dislocation density in region II gradually slows down as the system enters region III, but rather that the increase in dislocation density suddenly becomes slower when it reaches a certain value ($\rho_{II} = 9.1 \times 10^{14}$ m^{-2}). The significance of parameter ρ_{II} will be discussed later. Unlike in region II, the dislocation density slowly changes in region III. Next, the dislocation density suddenly decreases with unloading associated with fracture. This decrease indicates the occurrence of region IV. Since the time resolution of the experiment is 2 s, the dislocation density decreases to a quarter of that value during deformation

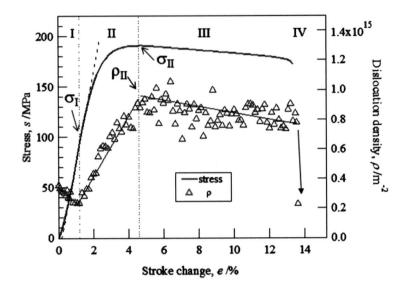

Fig. 9.1 Change in the dislocation density with the nominal strain for ARBed Aluminum during the in situ XRD measurement. Reproduced from (Adachi et al. 2020) and (Adachi et al. 2021) by permission of The Japan Institute of Light Metals

in less than 2 s, reaching the same level as that before the deformation. This is likely because, in UFG materials with a high grain boundary density, the boundary always exists close to the intra-grain dislocations, which leads to dislocations being annihilated, with the grain boundary acting as the sink in the unloading (Williamson and Smallman 1954). This means that in UFG materials, the dislocation substructure is quite different during deformation and after unloading, and that it is difficult to observe the structure during deformation by studying the dislocation substructure after unloading using an electron microscope. This can lead to a misunderstanding that even at room temperature, the driving component of the deformation in UFG materials is not dislocation, but other plastons, such as boundary sliding.

Figure 9.2a shows the changes in the dislocation density during tensile deformation of a 2 N-aluminum CG material with a grain size of 20 μm, formed by annealing a UFG material. Figure 9.2b shows the enlarged view around the low-strain side. In a CG material, the dislocation density also changes through four regions. However, region I, the elastic deformation region, is very short, and dislocations start to increase once the stress reaches 15 MPa ($=\sigma_I$), whereby the system transitions to region II. The fact that $\sigma_I = 102$ MP for the UFG material with a grain size of 0.5 μm shows that the dislocation source is activated at very low stress, and dislocations start to increase. In region II, the dislocations rapidly increase, just as in UFG materials, and the increase in the dislocation density becomes slow when it reaches a certain value ($\rho_{II} = 1.57 \times 10^{14}$ m^{-2}). The value of ρ_{II} is about one-sixth that of the UFG material, which is very small, and the system quickly reaches ρ_{II} after entering region II, so region II is barely observed. In region III, the dislocation density increases more

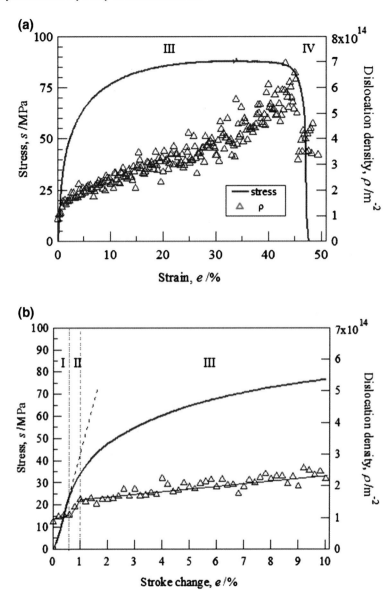

Fig. 9.2 Change in the dislocation density with the nominal strain for coarse-grained Al during the in situ XRD measurement. Reproduced from [22] and [39] by permission of The Japan Institute of Light Metals

slowly than that in region II and achieves a value of ~6 × 10^{14} m^{-2} right before the fracture. Next, as in UFG materials, the dislocation density increases more slowly with unloading associated with the fracture. However, the reduced amount is about half the increase in region III, and the dislocation density is much higher than that before the tension occurs. This is because the low grain boundary density in the CG material leads to little dynamic recovery. This shows that for the CG material, it is more or less possible to analogize the dislocation substructure during deformation, based on the dislocation substructure observed after unloading.

In the UFG material, the dislocation density rapidly increases in region II and does not increase much in region III, and the flow stress also does not increase. In contrast, for the CG material, the dislocation density significantly increases in region III, leading to an increase in the flow stress. Dislocations arise from the dislocation source in region III of the UFG material, but dynamic recovery, in which annihilation of the grain boundary as the sink occurs, progresses equally as fast, resulting in their balancing each other. As a result, the dislocation density hardly increases in region III, and the flow stress does not increase at all. Therefore, there is no work hardening in region III in the UFG material, which causes plastic instability and low ductility. On the other hand, in the CG material, there is little dynamic recovery due to the low grain boundary density, so the dislocation density significantly increases in region III and work hardening occurs, resulting in high ductility.

Next, a Ni with FCC structure is used to form even finer crystal grains because aluminum has a low melting point and low stacking fault energy, which leads to dynamic recovery during plastic deformation processing, and because it is difficult to reduce the diameters of the crystal grains of 2 N-aluminum via severe plastic deformation processing. Figure 9.3 shows the changes in dislocation density during tensile deformation in UFG nickel with a crystal grain size of 270 nm formed using ARB processing. The dislocation density of the UFG nickel also transitions through four regions, where in situ XRD measurements can be used to obtain the values of σ_I, ρ_{II}, and σ_{II}. For this material, the value of ρ_{II} is as high as 1.6 × 10^{15} m^{-2}, which is 1.8 times higher than that of the UFG aluminum with a particle diameter of 500 nm. Another difference between this material and the UFG aluminum with a grain diameter of 500 nm is that the dislocation density gradually increases in region III, which is likely to be because of nickel having a higher melting point than aluminum, and thus a lower stacking fault energy, resulting in slow dynamic recovery. However, in region IV, the unloading associated with fracture causes the dislocation density to instantaneously drop to a value close to that observed before the application of tension, and the dislocation substructure during the deformation of a UFG material is significantly different from that after unloading.

It is difficult to obtain finer nanocrystal grains using top-down methods such as high-strain processing (Dao et al. 2007; Liao et al. 2006), and thus a bottom-up method is required. Representative bottom-up strategies include electrolytic deposition, amorphous crystallization, and nanopowder solidification molding methods, which can produce a material with crystal grains of a single nanometer to a few tens of nanometers, sizes unobtainable via severe plastic deformation processing. Aluminum is a base metal, and a solution cannot be used as the electrolytic bath in

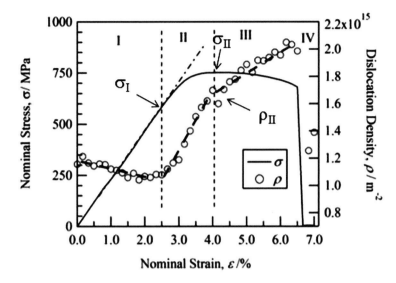

Fig. 9.3 Change in the dislocation density with the nominal strain for ARBed Ni during the in situ XRD measurement. Reproduced from (Adachi et al. 2016) by permission of The Japan Institute of Metals and Materials

the electrolytic deposition method, so the electrolytic deposition method is instead used to produce a nanocrystalline nickel material. A Watt bath is used with nickel sulfate hexahydrate to produce nickel nanocrystals with a crystal grain size of 50 nm (Schuh et al. 2002). Figure 9.4 shows the changes in dislocation density during tensile deformation for the nickel nanocrystals (Adachi et al. 2016). Like the CG and UFG materials, there are four regions in the dislocation density profile of the nickel nanocrystals (NC), where region II with a rapid increase in the dislocation density is extremely long and the dislocation density when moving between regions II and III has a ρ_{II} value of approximately 1.15×10^{16} m^{-2}, which is quite large compared to those of UFG nickel and CG aluminum. Moreover, as with UFG material, in region IV, unloading associated with fractures causes the dislocation density to decrease to a level observed before deformation, indicating that the dislocation substructure of the NC grains after unloading is significantly different from that during deformation.

In all of the in situ XRD measurements of materials, ranging from the coarse-grain aluminum with a particle size of 20 μm to the nanocrystalline nickel with a particle size of 50 nm, as described above, the dislocation density passes through four regions and the instantaneous decrease at the time of unloading associated with fracture becomes apparent as the grain size decreases. Let us consider what is signified by the ρ_{II} value required by in situ XRD measurement. Figure 9.5 shows the change in ρ_{II}, obtained via XRD measurements, of 2 N-Al with changing grain size, where the values for pure Ni are also assembled and shown. For grain sizes larger than 3 μm, ρ_{II} is almost constant at around 10^{14} m^{-2}, but when the grain size is less than 3 μm, ρ_{II} can be understood to be more or less proportional to the inverse

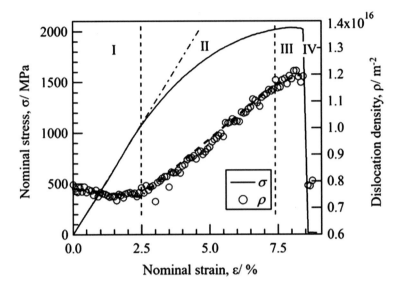

Fig. 9.4 Change in the dislocation density with the nominal strain for electrodeposited Ni during the in situ XRD measurement. Reproduced from (Adachi et al. 2016) by permission of The Japan Institute of Metals and Materials

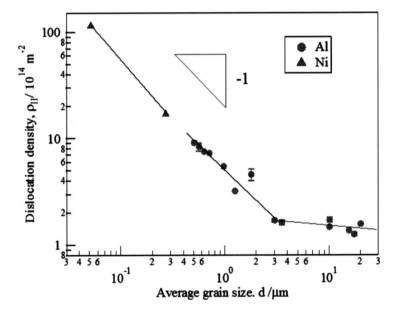

Fig. 9.5 The ρ_{II} as a function of grain size for pure Ni alloys and pure aluminum alloys. Reproduced from (Adachi et al. 2020) and (Adachi et al. 2021) by permission of The Japan Institute of Light Metals

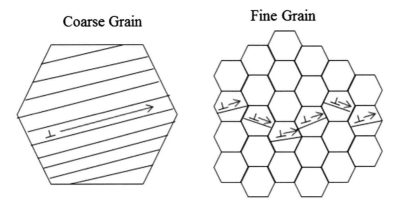

Fig. 9.6 Schematic illustration showing the change in mean free path of dislocation due to grain size. Reproduced from (Adachi et al. 2020) and (Adachi et al. 2021) by permission of The Japan Institute of Light Metals

of the grain size. This can be explained as follows. The relationship between the dislocation density and shear deformation, γ, per unit time by plastic deformation is

$$\gamma = \rho bx, \tag{9.3}$$

where b is the magnitude of Burger's vector and x is the mean free path. As shown in Fig. 9.6, x is large in the CG material and decreases in order for the increase in grain boundary density, which interferes with the dislocation motion, along with grain refinement. In other words, the dislocation density for plastic deformation increases as a result of grain refinement, and assuming that x is proportional to the particle size, the dislocation density is inversely proportional to the grain size, thus explaining the result in Fig. 9.5. The parameter x does not increase indefinitely with increasing grain size, but if the speed of dislocation motion is found to have an upper limit, then x likewise has an upper limit. Thus, there is a lower limit for ρ, which seems to be reached for particle sizes larger than 3 μm. Therefore, ρ_{II} is the least dislocation density necessary for deformation to occur only due to plastic deformation, and region II can be said to expand rapidly until this dislocation density is reached. Once it reaches ρ_{II}, dislocation no longer needs to increase so rapidly, so its growth slows, as does the growth speed of the dislocations, and the system transitions to region III. In addition, in region II, deformation cannot solely be achieved via plastic deformation, and elastic deformation makes up for the deficit, so the stress increases corresponding to elastic deformation. In other words, both regions II and III are plastic deformation regions. However, note that there is elastic deformation in region III as well, due to an increase in the flow stress caused by an increase in the dislocation density. For coarse crystal grains, the value of ρ_{II} is very small and is achieved rather quickly, leading to a short region II, and for fine grains, ρ_{II} is large, leading to a longer region II. Therefore, while region II is not so significant for CG materials, it needs

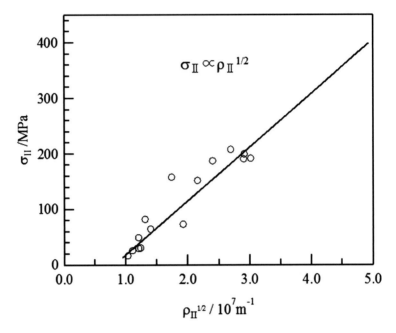

Fig. 9.7 The σ_{II} as a function of square root of ρ_{II} for pure aluminum alloys. Reproduced from (Adachi et al. 2020) and (Adachi et al. 2021) by permission of The Japan Institute of Light Metals

to be considered to understand the mechanical characteristics of UFG materials. For example, in UFG materials, a low initial dislocation density due to annealing leads to a small extent of plastic deformation in region II compared to a case with larger initial dislocations as a result of processing, resulting in a large elastic deformation and stress. However, the flow stress in region III depends on the dislocation density at the time and not on the initial dislocation density. Therefore, in UFG materials with low initial dislocation density, the stress in region II is greater than that in region III, making a yield point drop more probable.

Next, the relationship between the ρ_{II} of 2 N-aluminum with various-sized crystal grains and the stress σ_{II} is examined, as shown in Fig. 9.7. σII is proportional to the square root of the dislocation density, satisfying the Taylor relationship. In other words, the elementary process of plastic deformation involves dislocations cutting through Hayashi dislocations and plastic deformation progressing through dislocations in UFG material with a grain size of 500 nm. In addition, as shown in Fig. 9.5, the 50 nm nickel nanocrystals exhibit the same trend as that of aluminum with grain sizes of 500 nm to 20 μm, suggesting that plastic deformation progresses through dislocations up to a grain size of 50 nm.

It is difficult to obtain nanocrystalline materials with a crystal grain size of 50 nm or lower from pure metals by using an electrolytic deposition method; thus, alloying is required. Here, a nanocrystalline Ni–W alloy is developed in an electrolytic bath containing nickel sulfate hexahydrate and sodium tungstate (Nakayama et al. 2016;

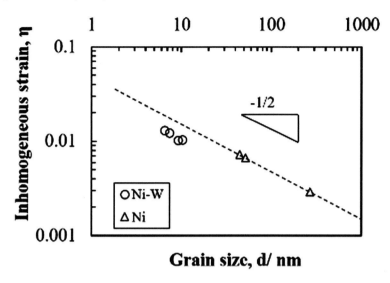

Fig. 9.8 The inhomogeneous strains as a function of grain size for NC-Ni and Ni-W alloys

Schuh et al. 2002; Yamasaki 2000, 1999). It is possible to change the W content by changing the ratios of nickel sulfate hexahydrate and sodium tungstate, and in this way, four alloys can be obtained: Ni-4.0at.%W, Ni-5.3at.%W, Ni-8.3at.%W, and Ni-9.8at.%W, with crystal grain sizes of 10.3, 9.2, 7.4, and 6.5 nm, respectively, and it is evident that the increase in W content decreases the crystal grain size. Figure 9.8 shows the changes in inhomogeneous strain, ε, during tensile deformation of the Ni-8.3at.%W alloy, which is a material with a grain size of a single nanometer. Assuming that this inhomogeneous strain occurs due to dislocations, the dislocation density can be obtained by substituting the inhomogeneous strain into Eq. (9.1). However, here the inhomogeneous strain is shown, which does not increase up to 2% macrostrain, indicating that the system is in region I, where only elastic deformation occurs. After that, the system enters region II, where the inhomogeneous strain increases linearly up to 5% macrostrain. At 5% macrostrain and above, changes in the inhomogeneous strain decelerate, indicating that the system is in region III. Based on the above observations, plastic deformation clearly occurs due to plastrons which generate inhomogeneous strain.

Denoting the inhomogeneous strain between regions II and III by ε_{II}, Fig. 9.8 shows the changes in ε_{II} due to the changes in the crystal grain sizes in the nanocrystalline Ni–W alloy. The figure also shows the ε_{II} values for nanocrystalline nickel obtained by the electrolytic deposition method and the UFG nickel obtained by the ARB method. As is clear from Fig. 9.5, ρ_{II} is proportional to the inverse of the grain size for pure nickel, so ε_{II} is proportional to the inverse of the square root of the grain size, as in Eq. (9.1). In Fig. 9.8, the ε_{II} value of the nanocrystalline Ni–W alloy is around 30–40% smaller than the dotted line with a slope value of $-1/2$ for pure nickel. Assuming that the plastic deformation progresses through dislocations, it can be surmised that the dislocation density from Eq. (9.1) is as low as 1/3–1/2 of the extrapolated values and that the deformation in the nanocrystalline Ni–W alloys with grain sizes of 6.5–10.3 nm does not progress through dislocations. If the plastic deformation in the nanocrystalline Ni–W alloys progresses through dislocations, the inhomogeneous strain is predicted by the extrapolation and the value of the dislocation density ρ_{II} necessary for plastic deformation to occur is almost 10^{17} m^{-2}. In this case, the strain energy in the grains becomes extremely high, making it likely that the plastic deformation progresses through a plaston with a lower inhomogeneous strain than dislocation.

What is the plaston in the nanocrystalline Ni–W alloy? As suggested by some molecular dynamics calculations, it is thought that partial dislocations have been activated (Yamasaki 1999; Yamakov et al. 2001). As is apparent from Eq. (9.1), the inhomogeneous strain is proportional to the square of the Burgers vector. For this reason, the activity of partial dislocations having a small Burgers vector suppresses the increase in the strain energy inside the grains, so the plaston in the single nanometer grain material is equivalent to a partial dislocation. In contrast, even though there are complete dislocation activities in materials ranging from CG aluminum to NC nickel, the total strain energy is relatively low, so the plastons are perfect dislocations. It has been reported that the addition of W to pure Ni reduces the stacking fault energy, and that the value for the Ni-10at.% W alloy is half that of pure Ni. In the Ni-10at.%W alloy (Suto and Kuniaki 1971), when the edge dislocations, which are complete dislocations, decompose into partial dislocations, the ditch of the lamination defect becomes ~ 5–6 nm. This is close to the crystal grain size, which makes it easier for partial dislocations to occur in the Ni–W alloy.

Figure 9.9 shows the changes in the 0.2% proof stress upon a change in the initial strain rate during tensile deformation in nanocrystalline nickel with a grain size of 50 nm and the Ni-5.3at.%W alloy with a grain size of 9.2 nm. The values of the strain rate sensitivity index, m, are found to be 0.036 and 0.026, respectively. It is known that an m value of greater than 0.3 occurs during boundary sliding and that of 1 occurs during Coble creep (Coble 1963). The m values are much lower than either of the above two values, which indicates that for materials with a single nanometer grain size at room temperature, the deformation does not progress through boundary sliding or creep, which is in good agreement with the in situ XRD results.

In coarse-grain materials, the Frank–Read source in the grains is the main dislocation source, but as the crystal grain becomes finer, the stress for generating dislocation from the Frank–Read source gradually increases. The dislocations bow out from the source, and since the stress required for increasing dislocations is inversely proportional to the source length, sources with longer source lengths can increase dislocations with lower stress. However, dislocations cannot bow out into the grains unless the source length is about a third of the grain size or smaller, and since the stress is inversely proportional to the grain size, the stress increases at an accelerated rate as the crystal grain size becomes finer. In UFG materials, the dislocation source shits at to the grain boundaries because the dislocation generation stress decreases (Kato et al. 2008; Kato 2009). The in situ XRD measurements mentioned above show that intra-grain dislocations are annihilated with the grain boundary because

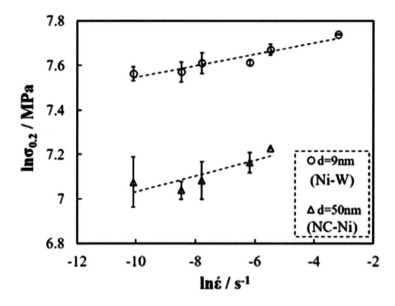

Fig. 9.9 Relationship between 0.2% proof stress and initial strain rate for NC-Ni and Ni-W alloys

of the sink occurring during unloading associated with fracture. From this, we can easily surmise that the grain boundary becomes a dislocation source. However, the grain boundary structure changes to a stable structure along with the dislocations emission, so it can be predicted that the grain boundary cannot emit dislocations without limit. In nanocrystalline nickel with a grain size of 50 nm formed by an electrolytic deposition method and the Ni–W alloy with a single nanometer grain size, the crystal grain size hardly changes due to the materials being kept at room temperature for a long time or being exposed to low-temperature annealing at around 373 K, but plastic elongation dramatically decreases. This suggests that, while the grain boundary of the nanocrystalline material developed by the electrolytic deposition method is at non-equilibrium and has a high potential for emitting dislocations, the potential decreases as the grain boundary structure becomes stable due to low-temperature annealing. How, then, can the dislocation release potential of the grain boundary be improved?

It is surmised that there is a limit to the extent of emit from the grain boundary in nanocrystalline materials, which is one of the causes for the low ductility of nanocrystalline materials. Attempts have been made to improve ductility by improving the dislocation emitted from the grain boundary. In other words, it is thought that the dislocation release potential can be improved by creating two phases of nanocrystalline and amorphous states, thereby placing a wide amorphous phase on the grain boundary of nanocrystals (Nakayama et al. 2016). It was previously mentioned that the W content in the Ni–W alloy can be changed by changing the ratios of nickel

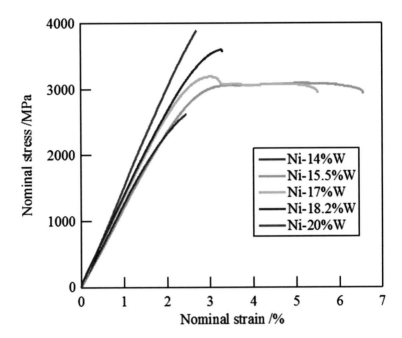

Fig. 9.10 Nominal stress–strain curves for Ni-W alloys with W contents between 14 and 20 at. %. Reprinted with permission from (Nakayama et al. 2016), Copyright 2016, American Scientific Publishers

sulfate hexahydrate and sodium tungstate in the electrolytic bath. The nanocrystalline single phase is maintained up to a W concentration of 14at%W, and the amorphous single phase takes over once it exceeds 20at%W. At ~14–20at%W, two phases, nanocrystalline and amorphous, can be obtained, and their proportion can be continuously changed by altering the W concentration. In addition, the grain size of the nanocrystals is an approximately constant value, at ~5 nm. Figure 9.10 shows the changes in the stress–strain curves due to changes in the W concentration in Ni–W alloys with nanocrystalline and amorphous phases. There is little ductility in the nanocrystalline single-phase Ni-14at%W and amorphous single-phase Ni-20at% alloys, but ductility is obtained in materials with nanocrystalline and amorphous phases. The Ni-17at%W alloy has a tensile strength of 2.5 GPa and a stretch of 4%, showing high strength and ductility. This is likely because the many amorphous regions in the grain boundaries of the nanocrystals increase the dislocation emit from the grain boundary, maintaining the plastic deformation.

Figure 9.11 shows the results of in situ XRD measurements on the Ni-17at%W alloy, for which because it is a two-phase alloy with nanocrystalline and amorphous phases, only a very wide (111) diffraction peak can be observed. This shows the change in the full width at half maximum (FWHM) of the (111) diffraction peak. It is evident that the FWHM suddenly decreases after the yield point, which implies that plastic deformation reduces the amorphous phase and increases the crystalline phase.

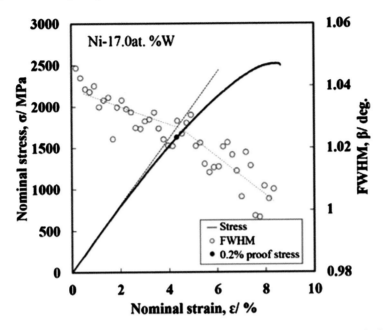

Fig. 9.11 Change in the FWHM of the Ni (111) peak as a function of nominal strain for Ni-12.7at.%W alloy during in situ XRD. Reprinted with permission from (Nakayama et al. 2016), Copyright 2016, American Scientific Publishers

An electron microscopy observation of the structure under tensile deformation with 5% plastic strain reveals that the nanocrystalline grain size increases, compared to that before the application of tension, by 1.1 nm. This suggests that the amorphous phases at the boundary of the nanocrystalline and amorphous phases become crystallized due to plastic deformation. In other words, as partial dislocations arise from the nanocrystalline and amorphous boundary during plastic deformation, the amorphous phase at the boundary becomes stable and changes its structure to FCC (Swygenhoven et al. 2002), upon which it is thought that the nanocrystals grow as a result.

References

Adachi H, Karamatsu Y, Nakayama S, Miyazawa T, Sato M, Yamasaki T (2016) Mater Trans 57:1447–1453
Adachi H, Miyajima Y, Sato M, Tsuji N (2015) Mater Trans 56:671–675
Adachi H, Mizowaki H, Hirata M, Okai D, Nakanishi H (2021) Mater Trans 62(1), in print
Adachi H, Mizowaki H, Hirata M, Okai D, Nakanishi H (2020) J JILM 70:274–280
Bailey JE, Hirsch PB (1960) Philos Mag 5:485–497
Budrovic Z, Swygenhoven HV, Derlet PM, Petegem SV, Schmitt B (2004) Science 304:273–276
Chen M, Ma E, Hemker KJ, Sheng H, Wang Y, Cheng X (2003) Science 300:1275–1277
Coble RL (1963) J Appl Phys 34:1679–1682

Dao M, Lu L, Asaro RJ, De Hosson JTM, Ma E (2007) Acta Mater 55:4041–4065
Ferrasse S (1997) Metall Mat Trans A 28:1047–1057
Fleischer RL (1963) Acta Metall 11:203–209
Gerold V, Harberkorn H (1966) Phys Stat Sol 16:675–685
Hall EQ (1951) Proc Phys Soc London B 64:747–751
Horita Z, Langdon TG (2005) Mater Sci Eng A 410–411:422–425
Huang X, Hansen N, Tsuji N (2006) Science 312:249–251
Huang X, Tsuji N, Hansen N, Minamino Y (2003) Mater Sci Eng A 340:265–271
Kamikawa N, Tsuji N, Saito Y (2003) Tetsu-to-Hagane 89:273–280
Kamikawa N, Huang X, Tsuji N, Hansen N (2009) Acta Mater 57:4198–4208
Kato M (2009) Mater Sci Eng A 516:276–282
Kato M, Fujii T, Onaka S (2008) Mater Trans 49:1278–1283
Liao XZ, Kilmametov AR, Valev RZ, Gao H, Li X, Mukherjee AK, Bingert JF, Zhu YT (2006)
    Appl Phys Lett 88:021909
Lu L, Schwaiger R, Shan ZW, Dan M, Lu K, Suresh S (2005) Acta Mater 53:2169–2179
Miyajima Y, Okubo S, Miyazawa T, Adachi H, Fujii T (2016) Philos Mag Lett 96:294–304
Mompiou F, Cailard D, Legros M, Mughrabi H (2012) Acta Mater. Col 60:3402–3414
Nakayama S, Adachi H, Nabeshima T, Miyazawa T, Yamasaki T (2016) Sci Adv Mater 8:2082–2088
Petch NJ (1953) J Iron Steel Inst 174:25–28
Schuh CA, Nieh TG, Yamasaki T (2002) Scripta Mater 46:735–740
Shen YF, Lu L, Lu QH, Jin ZH, Lu K (2005) Scripta Mater 52:989–994
Suto H, Kuniaki K (1971) J Jpn Inst Met 35:231–237
Swygenhoven HV, Spaczer M, Caro A (1999) Acta Mater 47:3117–3126
Swygenhoven HV, Derlet PM, Hasnaoui A (2002) Phys Rev B 66:024101
Tsuji N, Ito Y, Saito Y, Minamino Y (2002) Scripta Mater 47:893–899
Valiev RZ, Islamgaliev RK, Alexandrov IV (2000) Rrog Mater Sci 45:103–189
Williamson GK, Hall WH (1953) Acta Metall 1:23–31
Williamson GK, Smallman RE (1954) Acta Crystallogr 7:574–581
Williamson GK, Smallman RE (1956) Philos Mag 8:34–45
Yamakov V, Wolf D, Salazar M, Phillpot SR, Gleiter H (2001) Acta Mater 49:2713–2722
Yamasaki T (1999) J Jpn Inst Met Mater 69:404–412
Yamasaki T (2000) Mater Phys Mech 1:127–132

Chapter 10
Microstructural Crack Tip Plasticity Controlling Small Fatigue Crack Growth

Motomichi Koyama, Hiroshi Noguchi, and Kaneaki Tsuzaki

10.1 Introduction: Small Crack Problem

It has been generally recognized that resistance to small fatigue crack growth dominates the fatigue life and strength in numerous metallic structural components. "Small (short) cracks"[1] have been classified as mechanically small cracks, microstructurally small cracks, physically small cracks, and chemically small cracks (Ritchie and Lankford 1986), the details of which are listed in Table 10.1. In this chapter, we note the growth of mechanically and microstructurally small cracks. We first indicate that the growth behaviors of mechanically and microstructurally small cracks are completely different from those of large cracks. Representative examples of the small crack growth characteristics are shown in Fig. 10.1. The small crack growth rate at a certain stress intensity factor range (ΔK) cannot be obtained through the extrapolation of a large crack growth rate curve in an identical material (Suresh and

---

[1]In this paper, short and small cracks, which have been used for two- and three-dimensional cracks, are not distinguished for discussion.

M. Koyama · K. Tsuzaki
Institute for Materials Research, Tohoku University, Katahira 2-1-1, Aoba-ku, Sendai 980-8577, Miyagi, Japan
e-mail: koyama@imr.tohoku.ac.jp

Center for Elements Strategy Initiative for Structural Materials (ESISM), Kyoto University, Yoshida-honmachi, Sakyo-ku, Kyoto 606-8501, Japan

H. Noguchi · K. Tsuzaki
Department of Mechanical Engineering, Kyushu University, Moto-oka 744, Nishi-ku, Fukuoka 819-0395, Japan
e-mail: nogu@mech.kyushu-u.ac.jp

K. Tsuzaki (✉)
Research Center for Structural Materials, National Institute for Materials Science, Sengen 1-2-1, Tsukuba 305-0047, Japan
e-mail: tsuzaki.kaneaki@nims.go.jp

© The Author(s) 2022
I. Tanaka et al. (eds.), *The Plaston Concept*,
https://doi.org/10.1007/978-981-16-7715-1_10

Table 10.1 Definitions of small fatigue cracks (Ritchie and Lankford 1986)

Small crack	Characteristics	Criterion for determining a critical crack length	References
Microstructurally small	Microstructure-dependent Probabilistic crack growth	Critical microstructure size such, e.g. grain size	(Goto 1994; Omura et al. 2017; Koyama et al. 2017a; Chowdhury and Sehitoglu 2016)
Mechanically small	ΔK is unavailable	Plastic zone size	(Hironobu et al. 1992; Li et al. 2017a)
Physically small	Simply small, showing non-steady state crack growth	Engineering crack length e.g., < 0.5–1.0 mm	(Suresh and Ritchie 1984; Fukumura et al. 2017)
Chemically small	Chemical reaction on crack surfaces significantly acts	Up to 10 mm, depending on reaction kinetics and frequency	(Gangloff 1985)

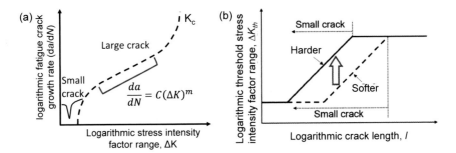

Fig. 10.1 Characteristics of small fatigue crack growth: (**a**) ΔK dependence of crack growth rate and (**b**) hardness sensitivity of ΔK_{th} with different initial crack lengths (Zhang et al. 2019). K_c: fracture toughness. " Reproduced with permission from Metall. Mater. Trans. A, 50A, 1142 (2019). Copyright 2019, The Minerals, Metals & Materials Society and ASM International"

Ritchie 1984) (Fig. 10.1(a)). Furthermore, the threshold stress intensity factor range (ΔK_{th}) for small crack growth is hardness-sensitive, unlike in the case of large cracks (Tanaka et al. 1981; Fukumura et al. 2015) (Fig. 10.1(b)). Therefore, we need specific mechanical and microstructural design strategies to improve the resistance to small fatigue crack growth, which differs from that for large crack growth resistance.

Small crack growth occurs through two routes: (A) dislocation–emission-induced crack tip deformation and (B) damage accumulation along a microstructural feature near a crack tip. Specifically, in case (A), when a crack tip opens, a dislocation emission occurs at the crack tip. Such a dislocation emission allows for a plastic crack opening, which advances the crack tip position (Fig. 10.2(a)). Even after a crack closes during an unloading and compression process, the crack tip position does not move backward. Hence, one cycle of crack opening and closing increases the

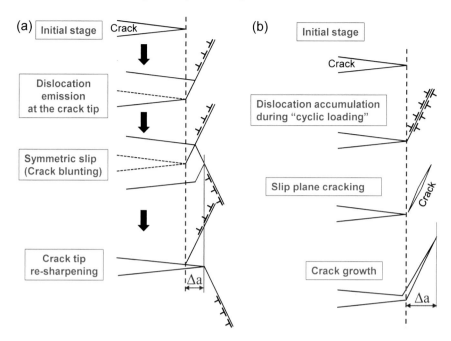

Fig. 10.2 Schematics of fatigue crack growth mechanisms driven by (**a**) dislocation emission at a crack tip and (**b**) damage accumulation near a crack tip

crack length, the degree of which is geometrically determined based on the number of emitted dislocations. In this regime, the lowest crack growth rate is an atomistic scale corresponding to one dislocation emission from the crack tip. In case (B), the role of the dislocations changes, namely, a local accumulation of dislocations causes micro-void or micro-crack formations near and/or at a crack tip through an increase in the local stress, the formation of vacancies, and the formation of a persistent slip band, which is referred to as damage accumulation and evolution. A damage-driven fatigue crack formation causes a discontinuous crack growth, which is strongly dependent on the microstructure of the crack tip (Fig. 10.2(b)). As understood from these basic mechanisms, the primary key to controlling the fatigue crack growth is crack tip plasticity. More specifically, the factors affecting the crack growth rate are divided into (1) resistance to plastic deformation, (2) resistance to damage/crack formation, (3) the crack tip deformation geometry, and (4) the crack closure effect. The first factor is directly associated with resistance to dislocation nucleation at a crack tip and to dislocation motion near a crack tip. The second factor is related to how the dislocation motion results in damage, such as vacancies, micro-voids, or micro-cracks. The third factor indicates the crystallographic relationship between the crack growth rate and crack tip/wake deformation. The fourth factor is a decrease in the actual stress or stress intensity factor at a crack tip, which is associated with a plastic zone evolution that realizes a lower crack tip stress than the elastic solution estimated based on the remote stress and crack shape and/or size. Considering the plasticity-controlled

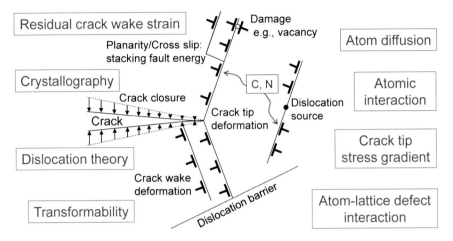

Fig. 10.3 Mechanical–metallurgical mechanism-based strategy for controlling small fatigue crack growth

factors, herein we introduce mechanical–metallurgical mechanism-based strategies for controlling small fatigue crack growth, as schematically shown in Fig. 10.3.

10.2 Grain Refinement: Characteristic Distributions of Dislocation Barrier and Source

First, we introduce the grain refinement effects on small crack growth, as well as a characteristic effect of ultrafine grains (UFG). As is well known, grain refinement increases the yield strength, which follows a Hall–Petch relationship (Tsuji et al. 2002). The enhancement of the yield strength and associated increase in hardness also increases the fatigue strength (Fig. 10.4(a)), which stems from the grain boundary acting as a dislocation barrier. Also note that the fatigue life of UFG steel at each stress amplitude is longer than that of coarse grained (CG) steel even after normalization by hardness (Fig. 10.4(b)), which implies an activation of extra factors. One possible factor is that grain boundaries can act as a dislocation source in ultrafine-grained materials owing to a lack of mobile dislocations in the grain interior. This effect alters the plastic strain gradient near the crack tip (Figs. 10.4(c, d)), which may affect the plasticity-induced crack closure (PICC) behavior. Moreover, the crack surface roughness, which also affects the crack growth resistance, is dependent on the grain size (Figs. 10.4(e, f)). In this section, we show the possible effects of a grain refinement on small fatigue crack growth in terms of the grain boundary roles as a dislocation source and barrier.[2]

---

[2] In this paper, we do not show the effect of grain size on crack roughness, because of the limitation of pages. Please see the references (Suresh and Ritchie 1982) Suresh S, Ritchie RO. A geometric

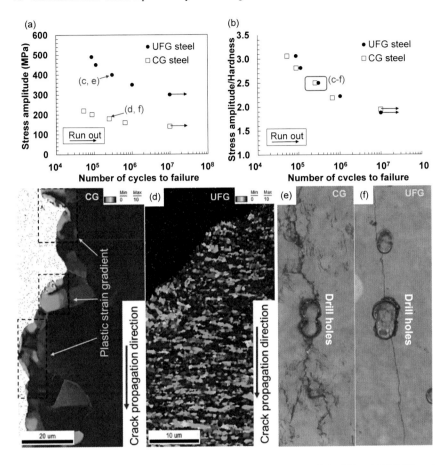

Fig. 10.4 Fatigue property and associated small fatigue crack growth in CG and UFG IF steels. The grain sizes of CG and UFG steels were 59 μm and 590 nm, respectively. (**a**) Stress amplitude–number of cycles to failure (S–N) diagram, and (**b**) corresponding hardness-normalized S–N curve. (**c, d**) Grain reference orientation deviation maps of CG and UFG steels. (**e, f**) Replica images showing a significant difference in crack roughness between CG and UFG steels (Lin et al. 2019). " Reproduced with permission from Int. J. Fatigue, 118, 117 (2019). Copyright 2018, Elsevier"

The crack tip plasticity interacts with the grain boundary, and its degree increases with a decrease in the distance between the crack tip and the grain boundary, which decelerates the crack growth. Hence, when the PICC effect is not considered, the

model for fatigue crack closure induced by fracture surface roughness. Metallurgical Transactions A. 1982;13:1627–31, (Niendorf et al. 2010) Niendorf T, Rubitschek F, Maier HJ, Canadinc D, Karaman I. On the fatigue crack growth–microstructure relationship in ultrafine-grained interstitial-free steel. Journal of Materials Science. 2010;45:4813–21, (Lin et al. 2019) Lin X, Koyama M, Gao S, Tsuji N, Tsuzaki K, Noguchi H. Resistance to mechanically small fatigue crack growth in ultrafine grained interstitial-free steel fabricated by accumulative roll-bonding. International Journal of Fatigue. 2019;118:117–25.

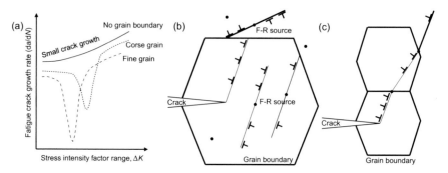

Fig. 10.5 Assumed grain size effect on small fatigue crack growth when the grain size is at the μm-scale or larger (Lankford 1982). (**a**) Schematics for dislocation emission, motion, and multiplication in (**b**) CG and (**c**) UFG metals. The black dots indicate a Frank–Read (F–R) source or grain boundary acting as a dislocation source

rate of crack growth decreases until the crack grows to the first grain boundary, and then increases after the crack penetrates the boundary. As schematically shown in Fig. 10.5(a), the grain refinement enhances the grain boundary effect (sharply decreasing the crack growth rate), although the effect only occurs for a very small crack length or low stress intensity factor range. Furthermore, UFG metals have been recognized to show plasticity-induced softening and subsequent plastic deformation propagation. When the grain size is large (CG metal), plastic deformation around the crack tip occurs through dislocation emissions from the crack tip and dislocation multiplication from the Frank–Read sources. Compared to this deformation in CG metals, UFG metals can show a different dislocation behavior owing to a low density of mobile dislocations in the grain interior (Huang et al. 2006). Specifically, once the crack tip plasticity initiates in UFG metals, grains surrounding the crack tip become highly distorted plastically because plasticity-induced softening occurs through mobile dislocation emissions and multiplication. The grain distortion associated with the accumulation and pile-up of geometrically necessary dislocations causes dislocation emissions from the grain boundaries to the neighboring grain, which results in the propagation of a grain-scale deformation. A chain reaction of the plastic deformation in UFG metals may decrease the gradient of plastic strain near the crack tip. Because a plastic strain gradient is one of the origins of PICC, the propagation of a deformation across the grain boundaries may decrease the contribution of PICC in UFG metals (Lin et al. 2019). Consequently, in terms of crack tip plasticity, grain refinement has an ambivalent role[2] in small crack growth because a grain boundary acting as dislocation barrier decelerates such growth, although when acting as a source it can accelerate the crack growth.

10.3   Plasticity-Induced Transformation: Thermodynamic-Based Design

10.3.1   Geometrical Effect on Crack Tip Deformation

In terms of crack tip plasticity, a martensitic transformation involving transformation dislocations can alter the crack growth behavior. Figure 10.6 shows an example of a transformation effect. The Fe–30Mn–6Al alloy in the figure shows no martensitic transformation and only dislocation slips. In contrast, the Fe–30Mn–6Si and Fe–30Mn–4Si–2Al alloys show a martensitic transformation from face-centered cubic (FCC: γ) to hexagonal close-packed (HCP: ε) structures. One difference between the Fe–30Mn–6Si and Fe–30Mn–4Si–2Al alloys is the plasticity of the HCP phase; namely, the HCP phase of the Fe–30Mn–6Si alloy shows brittle-like cracking, whereas that of the Fe–30Mn–4Si–2Al alloy is ductile (Nikulin et al. 2013). The crack growth rate of the Fe–30Mn–4Si–2Al alloy, which involves a γ-ε transformation, is significantly lower than that of the other alloys (Li et al. 2015; Ju et al. 2016), as shown in Fig. 10.6. More specifically, a "ductile" γ-ε martensitic transformation decelerates the crack growth. Considering these facts, two underlying mechanisms of crack growth deceleration can be proposed. The first factor is reversible γ ↔ ε transformation-induced plasticity (Sawaguchi et al. 2008, 2015), which can suppress the fatigue damage evolution associated with a vacancies and persistent slip bands near a crack tip (Ju et al. 2016). The second factor is a geometrical effect of the transformation dislocation motion forming the HCP structure (Ju et al. 2017).

Herein, we introduce a model for explaining the second factor, as schematically shown in Fig. 10.7. When a crack tip opens through a slip of perfect dislocations, repeated symmetric dislocation emissions occur at the crack tip, which causes fatigue

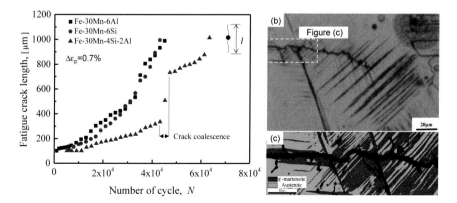

Fig. 10.6 Fatigue crack length plotted against number of cycles of three types of Fe–30Mn-based alloys (**a**). Optical image of Fe–30Mn–4Si–2Al specimen surface fatigued until 4.3×10^4 cycles (**b**). (**c**) Phase map of region highlighted in (**b**) (Ju et al. 2016). " Reproduced with permission from Acta Mater., 112, 326 (2016). Copyright 2016, Elsevier"

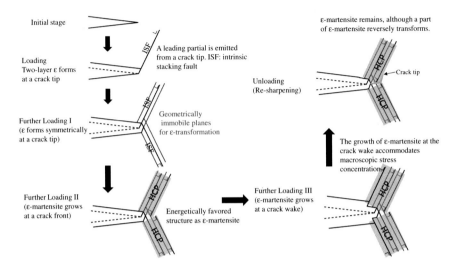

Fig. 10.7 Proposed model of a geometrical relationship between crack tip deformation and ε-martensitic transformation. The vertical direction corresponds to the loading direction (Ju et al. 2017). The red lines indicate the neighboring slip planes of the intrinsic stacking faults. "Reproduced with permission from Int. J. Fatigue, 103, 533 (2017). Copyright 2017, Elsevier"

crack growth for every cycle (Fig. 10.2(a)). In contrast, when the stacking fault energy is sufficiently low to induce a ε-martensitic transformation, the emission of a leading partial dislocation plays a role in the initial crack opening. To maintain the HCP structure, a second emission of a leading partial cannot occur on the identical slip plane. Furthermore, neighboring slip planes also cannot act as active planes for partial dislocation motions (if the partials move on the neighboring planes, the structure changes to an FCC twin.). Therefore, a second emission of leading partials must occur on the second-neighboring planes, which causes a crack wake deformation, and not a crack tip deformation. A crack wake deformation results in a crack opening, but does not contribute to Mode I crack growth. Because a portion of ε martensite can reversely transform during an unloading and compression loading process, the effect of the γ-ε martensitic transformation occurs repeatedly. This model suggests that a γ-ε martensitic transformation at a crack tip alters the ratio between the crack tip opening displacement and the rate of crack growth, which decelerates the crack growth, particularly in strain-controlled fatigue tests. In other words, the use of a group motion of dislocations decelerates the crack growth rate, the effect of which can be enhanced when the deformation-induced phase involves long-range periodic structures.

10.3.2 Transformation-Induced Hardening and Lattice Expansion

Figure 10.8(c) shows S–N diagrams of various types of austenitic steels with a small notch. All of the fatigue limits correspond to limits of fatigue crack non-propagation. As introduced later, Fe–23Mn–0.5C steel has been recognized to show a superior non-propagation limit compared to conventional austenitic steels, which can be used as a reference material for comparing the fatigue limit and strength. Note here that Fe–19Cr–8Ni–0.05C steel shows a martensitic transformation from FCC to a body-centered cubic structure (γ-α' martensitic transformation). The fatigue limit and strength of Fe–19Cr–8Ni–0.05C steel are higher than those of the other types of steels, including Fe–23Mn–0.5C steel, even after normalization based on the hardness determined prior to the fatigue tests. As shown in Fig. 10.9, the region surrounding the crack in Fe–19Cr–8Ni–0.05C steel is fully covered with α' martensite after the fatigue test even at the fatigue limit. In this context, the transformation-induced plasticity (TRIP) causes two phenomena that enhance the resistance to small fatigue

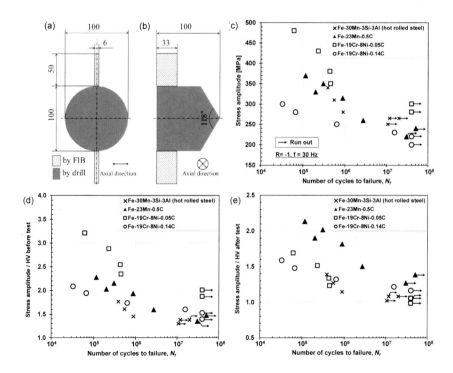

Fig. 10.8 Top and (b) side views of the micro-notch geometry (**a**). (**c**) S–N diagram of various types of austenitic steel. (**d, e**) S–N diagram normalized based on the Vickers hardness before and after the fatigue test at the fatigue limit (Nishikura et al. 2018). " Reproduced with permission from Int. J. Fatigue, 113, 359 (2018). Copyright 2018, Elsevier"

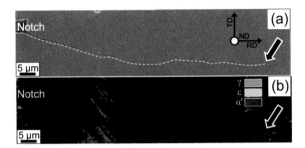

Fig. 10.9 SEM image (**a**) and (**b**) phase map of a location where a fatigue crack exists in Fe–19Cr–8Ni–0.05C steel after fatigue testing for 4.0×10^7 cycles at 300 MPa (fatigue limit) (Nishikura et al. 2018). The dashed line was drawn along the crack, and the arrows indicate a crack tip. " Reproduced with permission from Int. J. Fatigue, 113, 359 (2018). Copyright 2018, Elsevier"

crack growth: (1) work hardening (Ogawa et al. 2017, 2018) and (2) a transformation-induced crack closure (TICC) (Mei and Morris 1991; Mayer et al. 1995). The former is associated with a hard martensite phase and has been well recognized as having a TRIP effect during tensile tests, where the uniform elongation increases. The latter is associated with a lattice expansion arising from a close-packed structure (FCC) to a non-close-packed structure (BCC), which assists with the PICC. The TICC cannot act in the case of stress-controlled fatigue tests of Fe–19Cr–8Ni–0.05C steel because a γ-α' martensitic transformation occurs not only near a crack but also in non-cracked regions; namely, a homogeneous transformation cannot cause a crack tip constraint under a stress control condition (only the specimen volume increases). Therefore, the first fact, i.e., work hardening, is the primary factor triggering the superior resistance to fatigue crack growth in Fe–19–Cr–8Ni–0.05C steel. Correspondingly, when normalized based on the hardness evaluated after the fatigue tests at the fatigue limit, the fatigue limit of Fe–19Cr–8Ni–0.05C steel is lower than that of Fe–23Mn–0.5C steel (Figs. 10.8 (c) and (e)). By contrast, Fe–19Cr–8Ni–0.14C metastable austenitic steel, in which the extra carbon stabilizes the austenite (Ogawa et al. 2017), shows a lower fatigue limit than that of Fe–19Cr–8Ni–0.05C steel even when the fatigue limit is normalized based on the hardness prior to the fatigue test (Figs. 10.8 (c) and (d)), whereas the fatigue limit normalized by the hardness after the test in Fe–19Cr–8Ni–0.14C steel is higher (Fig. 10.8(e)). These facts indicate that the higher fatigue limit normalized based on the hardness after the test in Fe–19Cr–8Ni–0.14C steel is attributed to the TICC associated with local transformation around the fatigue crack, and not to the effect of the work hardening. These results suggest that, to maximize the work hardening and TICC, the austenite stability must be optimized to induce a martensitic transformation locally around the crack while avoiding such an occurrence in a non-cracked region of the specimen.

10.4   Dislocation Planarity: Stress Shielding and Mode II Crack Growth

Dislocation planarity is one of the most intrinsic characteristics controlling a long-range stress field and is enhanced by lowering the stacking fault energy and presence of short-range ordering, which increases the work hardening capability. Therefore, an enhancement of the dislocation planarity improves the tensile properties as long as no brittle cracking occurs; however, under fatigue, the dislocation planarity has both advantages and disadvantages regarding crack growth, which arise from the behaviors of the surface and interior of the specimen.

As a specific example, Fig. 10.10 shows a case of high-nitrogen austenitic steel with a chemical composition of Fe–25Cr–1 N that exhibits highly planar dislocation arrays owing to Cr–N coupling. As shown in Figs. 10.10 (a) and (b), the fatigue limit of high-nitrogen steel is significantly higher than that of Fe–Cr–Ni austenitic stainless steel without interstitials, although surprisingly, the steel has more potential in terms of small crack growth resistance. In fact, even at a stress amplitude above the fatigue limit, crack growth of the main crack stops temporarily; however, a surface subcrack initiation at the main crack tip and a subsequent coalescence restart the growth of the main crack. Hence, the ease of subcrack initiation on the surface deteriorates

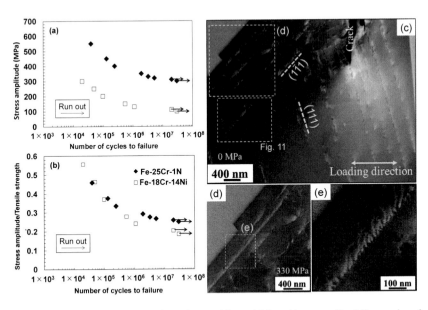

Fig. 10.10 S–N diagram of Fe–25Cr–1 N austenitic steel (**a**), and its normalized diagram based on the tensile strength (**b**) (Habib et al. 2017). ECC image showing crack tip microstructure after fatigue testing at 310 MPa until 2×10^7 cycles (**c**). ECC image showing dislocation multiplication on a specific slip plane after further loading to 330 MPa (**d**) and its magnification (**e**) (Habib et al. 2019a). "Reproduced with permissions from Int. J. Fatigue, 104, 158 (2017) Copyright 2017, Elsevier and Mater. Charact., 109,930 (2019) In press. Copyright 2019, Elsevier"

the resistance to fatigue crack growth, which is attributed to a slip localization. In high-nitrogen steel, the dislocation motion is atomistically limited on the specific slip plane, as shown in Figs. 10.10 (d) and (e). Because a specimen surface is basically under a plane stress condition, a highly planar slip near the surface causes a localized out-of-plane plastic deformation through a dislocation emission to the surface during the tensile loading process. The localized out-of-plane deformation results in a surface relief that acts as a stress concentration site, bringing about another planar-slip-driven out-of-plane slip deformation with a different sign of dislocation during the unloading/compression process. The repetition of this process causes a dislocation dipole formation and persistent slip bands, which assists in fatigue crack initiation and associated crack growth. However, when a dislocation planar array exists in the interior of a specimen, the dislocation motion stops at a barrier, such as Lomer–Cottrell sessile dislocations, which results in a strong pile-up stress (Fig. 10.11(a)). Such pile-up stress prevents further dislocation motion on the identical slip plane during tensile loading, as shown in Figs. 10.11 (b)–(d), the effect of which has been recognized to suppress crack growth along the slip plane (Awatani et al. 1979). Subsequently, the pile-up stress enables reversible motion of the dislocations and associated stacking faults during unloading/compression process (c.f. Figure 10.12), which also enhances resistance to crack growth. Moreover, the dislocation pile-up stress can act as a shield for the tensile stress at the crack tip, which suppresses crack opening; in particular, Mode I crack growth is also prevented by the enhancement of dislocation planarity. Therefore, if the surface crack initiation can be suppressed, the enhancement of dislocation planarity has a significant potential to improve the resistance to a small crack growth in austenitic steel with a high-nitrogen content.

Similar to this context, an equiatomic Fe–Cr–Ni–Mn–Co high-entropy alloy, which is a recently noted material, also shows a similar behavior (Suzuki et al. 2018). High-entropy alloy has been reported to exhibit a low stacking fault energy $(30 \pm 5 \text{ mJ/m}^2$ (Okamoto et al. 2016)) and a short-range ordering (Ding et al. 2018), thereby resulting in a highly planar slip in fatigued specimens. Correspondingly, once

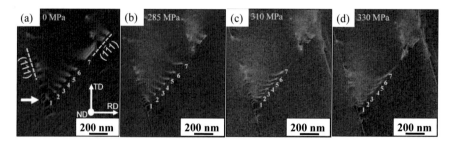

Fig. 10.11 ECC images during loading slightly far from the crack tip in the region outlined in Fig. 10.10(c) at (**a**) 0, (**b**) 285, (**c**) 310, and (**d**) 330 MPa. The white arrow in (a) indicates the location where multiple slips intersect each other (Habib et al. 2019a). " Reproduced with permission from Mater. Charact., 158, 109,930 (2019). Copyright 2019, Elsevier"

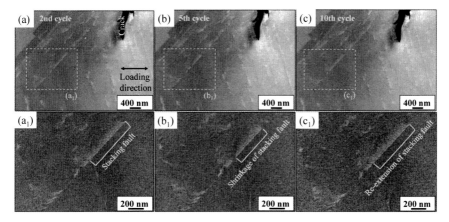

Fig. 10.12 ECC images during the cyclic loading after the 1-cycle in situ experiment of Figs. 10.10 and 10.11: (**a**) 2nd, (**b**) 5th, and (**c**) 10th cycles. (a_1), (b_1), and (c_1) Magnified images of the section outlined by the yellow line in (a), (b), and (c), respectively (Habib et al. 2019a). " Reproduced with permission from Mater. Charact., 158, 109,930 (2019). Copyright 2019, Elsevier"

the small crack growth stops, it restarts through a crack coalescence (Suzuki et al. 2018), which is the same characteristic found in high-nitrogen steel. As mentioned, lowering the stacking fault energy is essential to improving the tensile properties, and the short-range ordering is expected to be used for next-generation high-strength alloys such as high-nitrogen steel and high-entropy alloys. Therefore, controlling the dislocation planarity and its associated surface subcrack initiation is a key to realizing a drastic improvement in the fatigue crack growth resistance in high-tensile strength alloys.

10.5 Kinetic Effects of Solute Atoms on Crack Tip Plasticity

10.5.1 Strain-Age Hardening

The occurrence of strain-age hardening at a crack tip has been recognized to improve the non-propagation limit of fatigue cracking. For instance, solute Mg in Al alloys (Shikama et al. 2012; Zeng et al. 2012; Takahashi et al. 2015) and solute carbon in steel (Li et al. 2016, 2017b, 2018; Kishida et al. 2018), both of which trigger strain aging, significantly improve the ΔK_{th} of small crack growth. Figure 10.13 shows an example of the carbon effect on ΔK_{th} in steel. The occurrence of dynamic strain aging can be confirmed in terms of the stress–strain response under tension. Interstitial free (IF) steel, which does not contain solute carbon, shows a smooth stress–strain curve, regardless of the strain rate (Fig. 10.13(a)). In contrast, Fe–0.017C steel, in which all carbon exists in a solute form, shows a serrated flow, particularly at low

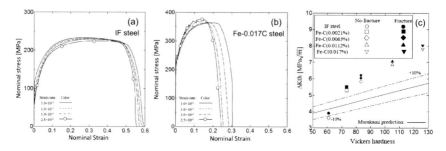

Fig. 10.13 Nominal stress–strain curves of (**a**) IF and (**b**) Fe-0.017C steel. (**c**) ΔK_{th} plotted against Vickers hardness (Li et al. 2017b). The dashed lines in (**c**) indicate a 10% error from the prediction using Murakami's equation. " Reproduced with permission from Int. J. Fatigue, 98, 101 (2017). Copyright 2017, Elsevier."

strain rates (Fig. 10.13(b)). The occurrence of a serrated flow is evidence of the contribution of dynamic strain aging to the stress–strain response. The degree of a serrated flow increases with an increase in the solute carbon from 0.002 to 0.017 mass% (Li et al. 2017b). Correspondingly, ΔK_{th} increases with an increase in the solute carbon concentration, as shown in Fig. 10.13(c). To explain the details shown in Fig. 10.13(a), herein we introduce an empirical relationship between the hardness and ΔK_{th} proposed by Murakami (Murakami 2019).

$$\Delta K_{th} = 3.3 \times 10^{-3}(HV + 120)\left(\sqrt{area}\right)^{1/3}, \tag{10.1}$$

where HV is the Vickers hardness, $\sqrt{area}$ is the square root of the projected area of the initial defects, such as pre-cracks or inclusions. The effectiveness of the Murakami equation has already been demonstrated for the ΔK_{th} predictions of carbon steel with different initial notch geometries, and the prediction error was determined to be less than 10% (Li et al. 2016; Murakami 2019). The solid and dotted lines in Fig. 10.13(c) indicate the predicted values and their 10% errors, respectively. The measured ΔK_{th} of IF steel is within the $\pm 10\%$ error band shown in Fig. 10.13(c), whereas the four types of Fe–C steel with different carbon contents show a higher ΔK_{th} than the predicted value of $+ 10\%$. Furthermore, the difference between the predicted and measured ΔK_{th} values increases with an increase in the carbon content. These facts indicate that the dynamic strain aging arising from solute carbon occurs particularly in the plastic zone, which increases the crack tip hardness during fatigue tests. As a result, the presence of solute carbon overcomes the conventionally known hardness–ΔK_{th} relation.

10.5.2 Effects of i–s Interaction

A problematic issue in utilizing the effect of strain-age hardening is a limitation of the solute diffusivity at room temperature. For example, normal austenitic steel does not show dynamic strain-age hardening even if the steel contains carbon or nitrogen in the solute because their diffusivity is insufficient for μm-scale atom motions. This problem can be solved using attractive interactions of interstitial–substitutional atoms (i–s). As introduced in Sect. 10.4, Cr–N interaction enhances the dislocation planarity, whereas another type of i–s interaction, i.e., Mn–C, assists strain-age hardening in austenitic steel at room temperature (Koyama et al. 2018a). Figure 10.14(a) shows an S–N diagram of Fe–23Mn–0.5C and Fe–30Mn–3Si–3Al austenitic steels. Both types of steels show deformation twinning and a dislocation glide, although only Fe–23Mn–0.5C steel demonstrates dynamic strain-age hardening even at room temperature. Fe–30Mn–3Si–3Al does not show non-propagating fatigue cracks, whereas Fe–23Mn–0.5C clearly does, as indicated in Figs. 10.14(b–d); namely, the Mn–C coupling improves the limit of fatigue crack non-propagation

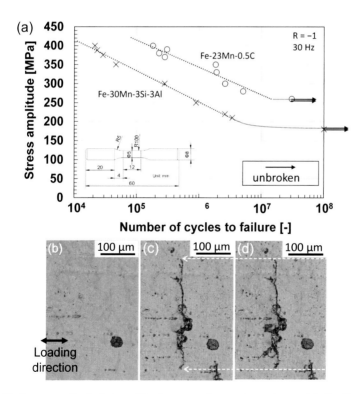

Fig. 10.14 Stress amplitude–fatigue life diagrams of smooth specimens (**a**). Replica images showing the non-propagation of a small crack in Fe–23Mn–0.5C at (**b**) 0, (**c**) 1×10^7, and (**d**) 3×10^7 cycles (Koyama et al. 2017b). " Reproduced with permission from Int. J. Fatigue, 94, 1 (2017) Copyright 2016, Elsevier."

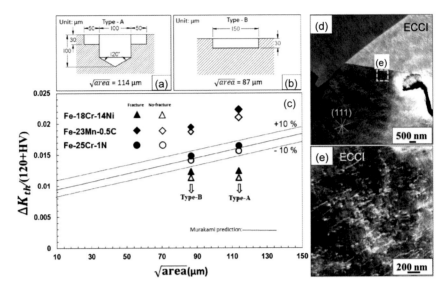

Fig. 10.15 Micro-notch geometries (**a**, **b**) and (**c**) associated ΔK_{th} data normalized using 120 + HV. (**d**) Non-propagating crack tip microstructure and (**e**) its magnification in Fe–23Mn–0.5C (Habib et al. 2019b). " Reproduced with permission from Metall. Mater. Trans. A, 50A, 426 (2019). Copyright 2018, The Minerals, Metals & Materials Society and ASM International"

through a triggering of the dynamic strain-age hardening. Evaluating the ΔK_{th} of Fe–23Mn–0.5C steel, the non-propagation limit of a fatigue crack is higher than that of high-nitrogen austenitic steel. According to the ECCI characterization (Figs. 10.15(d, e)), a high dislocation density was confirmed at the tip of the non-propagating fatigue crack. The high dislocation density increases the crack tip hardness, which thereby improves the non-propagation limit of the fatigue crack. Hence, the use of i-s attractive interactions for strain-age hardening can be applied as a new alloy design strategy for the development of materials with a high resistance to high-cycle fatigue.

10.6 Effect of Microstructural Hardness Heterogeneity: Discontinuous Crack Tip Plasticity

The final example of a mechanical–metallurgical mechanism-based strategy is the use of a microstructural hardness heterogeneity. In a general sense, the introduction of soft portions is speculated to decrease the resistance to fatigue crack propagation; however, in a controlled case, the soft portion can increase the resistance to fatigue crack propagation owing to an enhancement of PICC (Li et al. 2017a), an example of which is shown in Fig. 10.16. Figure 10.16(a) shows an RD-IPF map of hot-rolled Fe–30Mn–3Si–3Al steel. Because of the recrystallization during hot-rolling, the grain size distribution appears heterogeneously. Moreover, the degree of plastic

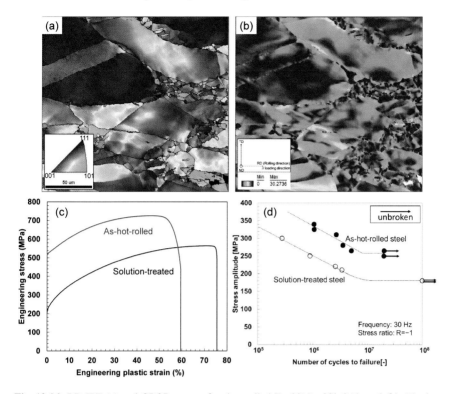

Fig. 10.16 RD-IPF (**a**) and GROD maps of as-hot-rolled Fe–30Mn–3Si–3Al steel (**b**). The hot-rolling temperature was 1,273 K. Engineering stress–strain (**c**) and stress amplitude–number of cycles to failure diagrams of the as-hot-rolled and solution-treated Fe–30Mn–3Si–3Al steel (**d**) (Koyama et al. 2018b). " Reproduced with permission from Int. J. Fatigue, 108, 18 (2018). Copyright 2017, Elsevier"

deformation is also heterogeneous (Fig. 10.16(b)). The partially fine grain and high dislocation density increase the yield and tensile strength, as shown in Fig. 10.16(c). The increase in tensile strength increases the fatigue limit (Fig. 10.16(d); however, the fatigue limits normalized by the respective tensile strength of the as-hot-rolled and solution-treated steel were 0.36 and 0.32, respectively. Furthermore, the as-hot-rolled steel showed a non-propagating fatigue crack, whereas the solution-treated steel did not. An image of the non-propagating fatigue crack is shown in Fig. 10.17(a), and the crack is presented in the grain interior (Fig. 10.17(b). For further characterization of the crack, the specimen surface was slightly polished mechanically, as shown in Figs. 10.17 (c) and (d), and the identical region surrounding the crack tip was analyzed using EBSD (Figs. 10.17(e, f)). The crack tip was observed in the grain interior, as indicated in Fig. 10.17(d), and the crack propagated from a highly distorted region to a less distorted region, shown in Fig. 10.17(f). In other words, the crack propagated from the soft to hard regions, and stopped within the grain interior. According to a mechanical simulation using a Dugdale model, the plasticity mismatch at the interface

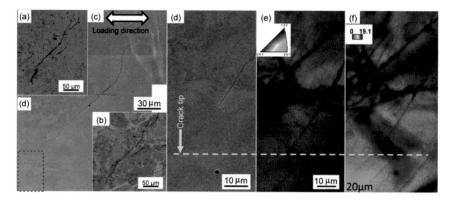

Fig. 10.17 A replica image (**a**) and corresponding optical micrograph of a chemically etched surface (**b**), which show a non-propagating crack in the as-hot-rolled Fe–30Mn–3Si–3Al steel. Overview of the fatigue crack after slight mechanical polishing (**c**). A magnified SEM image (**d**), and corresponding (**e**) RD-IPF and (**f**) GROD maps (Koyama et al. 2018b). " Reproduced with permission from Int. J. Fatigue, 108, 18 (2018). Copyright 2017, Elsevier"

of the soft and hard regions can enhance the effect of the PICC (Li et al. 2017a), which assists in the non-propagation of fatigue cracking. Therefore, even if the hardness or tensile strength is macroscopically identical, control of the microstructural hardness heterogeneity can improve resistance to fatigue crack growth.

10.7 Summary

In this chapter, we introduced mechanical–metallurgical mechanism-based strategies for improving the crack growth resistance in terms of the crack tip plasticity. Figure 10.18 schematically summarizes the factors that can enhance the resistance to small crack growth. Although respective factors were separately introduced, such factors interact with each other. For instance, the factors hardening the crack tip region definitely affect the behavior of a crack wake deformation. Therefore, thus far, a prediction of the small crack growth has been recognized to be a difficult issue. However, the coupling of multiple mechanisms can realize unexpected fatigue resistance through synergetic effects. In fact, we succeeded in presenting an outstanding fatigue resistance associated with the activation of multi-mechanisms in steel (Koyama et al. 2017c). We believe that there is a large frontier for fatigue-resistant material development on the design of a mechanical–metallurgical mechanism-based microstructure.

In addition, uncertain problems remain regarding small crack growth. Mode II type crack growth associated with damage (lattice defect) accumulation is one remaining unknown, but has frequently been observed in various types of metal including steel and titanium alloy (Habib et al. 2017; Maenosono et al. 2018, 2019; Mizumachi

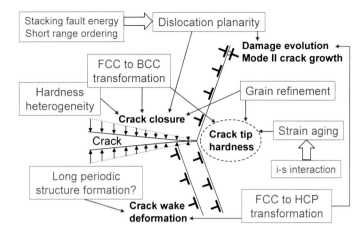

Fig. 10.18 Causes of the presented factors being affected by the fatigue crack growth

et al. 2019). The ease of Mode II crack growth affects the crack roughness (crack growth path), and discontinuous crack growth, and highly disturbs the mechanical laws of fatigue crack growth. The kinetic effects of solute atoms also still contain numerous uncertainties, which affect the temperature and frequency dependence of fatigue resistance. In fact, the importance of such kinetic effects has recently increased owing to the occurrence of hydrogen-accelerated fatigue crack growth in metal used for a hydrogen-related infrastructure (Murakami and Matsuoka 2010; Onishi et al. 2016). The problems of Mode II crack growth and kinetic effects of the solutes are clearly a mechanical–metallurgical issue, which is expected to be solved through the cooperation of mechanical engineers and metallurgists.

Acknowledgements This work was supported by Elements Strategy Initiative for Structural Materials (ESISM: JPMXP0112101000) of Ministry of Education, Culture, Sports, Science and Technology (MEXT) Japan, and JSPS KAKENHI (JP16H06365; JP20H02457).

References

Awatani J, Katagiri K, Koyanagi K (1979) A study on the effect of stacking fault energy on fatigue crack propagation as deduced from dislocation patterns. Metall Trans A 10:503–507

Chowdhury P, Sehitoglu H (2016) Mechanisms of fatigue crack growth–A critical digest of theoretical developments. Fatigue Fract Eng Mater Struct 39:652–674

Ding J, Yu Q, Asta M, Ritchie RO (2018) Tunable stacking fault energies by tailoring local chemical order in CrCoNi medium-entropy alloys. Proc Natl Acad Sci USA 115:8919–8924

Fukumura N, Suzuki T, Hamada S, Tsuzaki K, Noguchi H (2015) Mechanical examination of crack length dependency and material dependency on threshold stress intensity factor range with Dugdale model. Eng Fract Mech 135:168–186

Fukumura N, Li B, Koyama M, Suzuki T, Hamada S, Tsuzaki K et al (2017) Material property controlling non-propagating fatigue crack length of mechanically and physically short-crack based on Dugdale-model analysis. Theor Appl Fract Mech 90:193–202

Gangloff RP (1985) Crack size effects on the chemical driving force for aqueous corrosion fatigue. Metall Trans A 16:953

Goto M (1994) Statistical investigation of the behaviour of small cracks and fatigue life in carbon steels with different ferrite grain sizes. Fatigue Fract Eng Mater Struct 17:635–649

Habib K, Koyama M, Tsuchiyama T, Noguchi H (2017) Fatigue crack non-propagation assisted by nitrogen-enhanced dislocation planarity in austenitic stainless steels. Int J Fatigue 104:158–170

Habib K, Koyama M, Tsuchiyama T, Noguchi H (2019b) ECCI characterization of dislocation structures at a non-propagating fatigue crack tip: Toward understanding the effects of Mn-C and Cr-N couples on crack growth resistance. Metall Mater Trans A 50:426–435

Habib K, Koyama M, Tsuchiyama T, Noguchi H (2019) Dislocation motion at a fatigue crack tip in a high-nitrogen steel clarified through in situ electron channeling contrast imaging. Mater Charact 158:109930

Hironobu N, Masahiro G, Norio K (1992) A small-crack growth law and its related phenomena. Eng Fract Mech 41:499–513

Huang X, Hansen N, Tsuji N (2006) Hardening by annealing and softening by deformation in nanostructured metals. Science 312:249–251

Ju Y-B, Koyama M, Sawaguchi T, Tsuzaki K, Noguchi H (2016) In situ microscopic observations of low-cycle fatigue-crack propagation in high-Mn austenitic alloys with deformation-induced ε-martensitic transformation. Acta Mater 112:326–336

Ju Y-B, Koyama M, Sawaguchi T, Tsuzaki K, Noguchi H (2017) Effects of ε-martensitic transformation on crack tip deformation, plastic damage accumulation, and slip plane cracking associated with low-cycle fatigue crack growth. Int J Fatigue 103:533–545

Kishida K, Koyama M, Yoshimura N, Sakurada E, Yokoi T, Ushioda K et al (2018) Effect of Si on temperature dependence of non-propagation limit of small fatigue crack in a Fe-C alloy. Procedia Struct Integr 13:1032–1036

Koyama M, Li H, Hamano Y, Sawaguchi T, Tsuzaki K, Noguchi H (2017a) Mechanical-probabilistic evaluation of size effect of fatigue life using data obtained from single smooth specimen: An example using Fe-30Mn-4Si-2Al seismic damper alloy. Eng Fail Anal 72:34–47

Koyama M, Yamamura Y, Che R, Sawaguchi T, Tsuzaki K, Noguchi H (2017b) Comparative study on small fatigue crack propagation between Fe-30Mn-3Si-3Al and Fe-23Mn-0.5C twinning-induced plasticity steels: Aspects of non-propagation of small fatigue cracks. Int J Fatigue 94:1–5

Koyama M, Zhang Z, Wang M, Ponge D, Raabe D, Tsuzaki K et al (2017c) Bone-like crack resistance in hierarchical metastable nanolaminate steels. Science 355:1055–1057

Koyama M, Sawaguchi T, Tsuzaki K (2018a) Overview of dynamic strain aging and associated phenomena in Fe–Mn–C steels. ISIJ Int 58:1383–1395

Koyama M, Yamamura Y, Sawaguchi T, Tsuzaki K, Noguchi H (2018b) Microstructural hardness heterogeneity triggers fatigue crack non-propagation in as-hot-rolled Fe-30Mn-3Si-3Al twinning-induced plasticity steel. Int J Fatigue 108:18–24

Lankford J (1982) The growth of small fatigue cracks in 7075–T6 aluminum. Fatigue Fract Eng Mater Struct 5:233–248

Li H, Koyama M, Sawaguchi T, Tsuzaki K, Noguchi H (2015) Importance of crack-propagation-induced ε-martensite in strain-controlled low-cycle fatigue of high-Mn austenitic steel. Philos Mag Lett 95:303–311

Li B, Koyama M, Sakurada E, Yoshimura N, Ushioda K, Noguchi H (2016) Potential resistance to transgranular fatigue crack growth of Fe–C alloy with a supersaturated carbon clarified through FIB micro-notching technique. Int J Fatigue 87:1–5

Li B, Koyama M, Hamada S, Noguchi H (2017a) Threshold stress intensity factor range of a mechanically-long and microstructually-short crack perpendicular to an interface with plastic mismatch. Eng Fract Mech 182:287–302

Li B, Koyama M, Sakurada E, Yoshimura N, Ushioda K, Noguchi H (2017b) Underlying interstitial carbon concentration dependence of transgranular fatigue crack resistance in Fe-C ferritic steels: The kinetic effect viewpoint. Int J Fatigue 98:101–110

Li B, Koyama M, Sakurada E, Yoshimura N, Ushioda K, Noguchi H (2018) Temperature dependence of transgranular fatigue crack resistance in interstitial-free steel and Fe-C steels with supersaturated carbon: Effects of dynamic strain aging and dynamic precipitation. Int J Fatigue 110:1–9

Lin X, Koyama M, Gao S, Tsuji N, Tsuzaki K, Noguchi H (2019) Resistance to mechanically small fatigue crack growth in ultrafine grained interstitial-free steel fabricated by accumulative roll-bonding. Int J Fatigue 118:117–125

Maenosono A, Koyama M, Tanaka Y, Ri S, Wang Q, Noguchi H (2018) Crystallographic orientation-dependent growth mode of microstructurally fatigue small crack in a laminated Ti–6Al–4V alloy. Procedia Struct Integr 13:694–699

Maenosono A, Koyama M, Tanaka Y, Ri S, Wang Q, Noguchi H (2019) Crystallographic selection rule for the propagation mode of microstructurally small fatigue crack in a laminated Ti-6Al-4V alloy: Roles of basal and pyramidal slips. Int J Fatigue 128:105200

Mayer HR, Stanzl-Tschegg SE, Sawaki Y, Hühner M, Hornbogen E (1995) Influence of transformation-induced crack closure on slow fatigue crack growth udner variable amplitude loding. Fatigue Fract Eng Mater Struct 18:935–948

Mei Z, Morris JW (1991) Analysis of transformation-induced crack closure. Eng Fract Mech 39:569–573

Mizumachi S, Koyama M, Fukushima Y, Tsuzaki K (2019) Growth behavior of a mechanically long fatigue crack in an FeCrNiMnCo high entropy alloy: A comparison with an austenitic stainless steel. Tetsu-to-Hagane 105:215–221

Murakami Y (2019) Metal fatigue: Effects of small defects and nonmetallic inclusions. Academic Press, Cambridge, USA

Murakami Y, Matsuoka S (2010) Effect of hydrogen on fatigue crack growth of metals. Eng Fract Mech 77:1926–1940

Niendorf T, Rubitschek F, Maier HJ, Canadinc D, Karaman I (2010) On the fatigue crack growth–microstructure relationship in ultrafine-grained interstitial-free steel. J Mater Sci 45:4813–4821

Nikulin I, Sawaguchi T, Tsuzaki K (2013) Effect of alloying composition on low-cycle fatigue properties and microstructure of Fe–30Mn–(6–x)Si–xAl TRIP/TWIP alloys. Mater Sci Eng: A 587:192–200

Nishikura Y, Koyama M, Yamamura Y, Ogawa T, Tsuzaki K, Noguchi H (2018) Non-propagating fatigue cracks in austenitic steels with a micro-notch: Effects of dynamic strain aging, martensitic transformation, and microstructural hardness heterogeneity. Int J Fatigue 113:359–366

Ogawa T, Koyama M, Tasan CC, Tsuzaki K, Noguchi H (2017) Effects of martensitic transformability and dynamic strain age hardenability on plasticity in metastable austenitic steels containing carbon. J Mater Sci 52:7868–7882

Ogawa T, Koyama M, Nishikura Y, Tsuzaki K, Noguchi H (2018) Fatigue behavior of Fe-Cr-Ni-based metastable austenitic steels with an identical tensile strength and different solute carbon contents. ISIJ Int 58:1910–1919

Okamoto NL, Fujimoto S, Kambara Y, Kawamura M, Chen ZMT, Matsunoshita H et al (2016) Size effect, critical resolved shear stress, stacking fault energy, and solid solution strengthening in the CrMnFeCoNi high-entropy alloy. Sci Rep 6:35863

Omura T, Koyama M, Hamano Y, Tsuzaki K, Noguchi H (2017) Generalized evaluation method for determining transition crack length for microstructurally small to microstructurally large fatigue crack growth: Experimental definition, facilitation, and validation. Int J Fatigue 95:38–44

Onishi Y, Koyama M, Sasaki D, Noguchi H (2016) Characteristic fatigue crack growth behavior of low carbon steel under low-pressure hydrogen gas atmosphere in an ultra-low frequency. ISIJ Int 56:855–860

Ritchie RO, Lankford J (1986) Small fatigue cracks: A statement of the problem and potential solutions. Mater Sci Eng 84:11–16

Sawaguchi T, Bujoreanu L-G, Kikuchi T, Ogawa K, Koyama M, Murakami M (2008) Mechanism of reversible transformation-induced plasticity of Fe–Mn–Si shape memory alloys. Scr Mater 59:826–829

Sawaguchi T, Nikulin I, Ogawa K, Sekido K, Takamori S, Maruyama T et al (2015) Designing Fe–Mn–Si alloys with improved low-cycle fatigue lives. Scr Mater 99:49–52

Shikama T, Takahashi Y, Zeng L, Yoshihara S, Aiura T, Higashida K et al (2012) Distinct fatigue crack propagation limit of new precipitation-hardened aluminium alloy. Scr Mater 67:49–52

Suresh S, Ritchie RO (1982) A geometric model for fatigue crack closure induced by fracture surface roughness. Metall Trans A 13:1627–1631

Suresh S, Ritchie RO (1984) Propagation of short fatigue cracks. Int Metals Rev 29:445–475

Suzuki K, Koyama M, Noguchi H (2018) Small fatigue crack growth in a high entropy alloy. Procedia Struct Integr 13:1065–1070

Takahashi Y, Shikama T, Nakamichi R, Kawata Y, Kasagi N, Nishioka H et al (2015) Effect of additional magnesium on mechanical and high-cycle fatigue properties of 6061–T6 alloy. Mater Sci Eng: A 641:263–273

Tanaka K, Nakai Y, Yamashita M (1981) Fatigue growth threshold of small cracks. Int J Fract 17:519–533

Tsuji N, Ito Y, Saito Y, Minamino Y (2002) Strength and ductility of ultrafine grained aluminum and iron produced by ARB and annealing. Scr Mater 47:893–899

Zeng L, Shikama T, Takahashi Y, Yoshihara S, Aiura T, Noguchi H (2012) Fatigue limit of new precipitation-hardened aluminium alloy with distinct fatigue crack propagation limit. Int J Fatigue 44:32–40

Zhang Z, Koyama M, Wang M, Tasan CC, Noguchi H (2019) Fatigue resistance of laminated and non-laminated TRIP-maraging steels: Crack roughness vs. tensile strength. Metall Mater Trans A 50:1142–1145

Part IV
Design and Development of High Performance Structural Materials

Chapter 11
Designing High-Mn Steels

Takahiro Sawaguchi

11.1 Introduction

High-Mn austenitic steels have attractive mechanical and functional properties associated with characteristic plasticity mechanisms in the γ-austenite that has a face-centred cubic (FCC) structure. These include an extended dislocation glide, mechanical twinning, and mechanical martensitic transformation into ε-martensite with a hexagonal close-packed (HCP) structure and further to α'-martensite with a body-centred cubic or tetragonal (BCC/BCT) structure. Extensive research has been conducted on the deformation mechanism of high-Mn steels, particularly on the transformation- and twinning-induced plasticity (TRIP/TWIP) effects, which lead to the remarkable combination of high strength, ductility, and toughness.

Hereinafter, high-Mn austenitic steel is defined in a broad sense as the steel characterised by a high concentration of Mn, an initially fully austenitic phase state, and low-to-moderate stacking fault energies (SFEs). This category includes traditional Hadfield steels (Hadfield 1888), cryogenic high-Mn steels (Charles et al. 1981; Kim et al. 2015; Sohn et al. 2015), non-magnetic high-Mn steels, TRIP/TWIP steels (Grassel et al. 2000; Bouaziz et al. 2011; Cooman et al. 2018), high damping Fe–Mn alloys (Wang et al. 2019; Shin et al. 2017; Jee et al. 1997), and Fe–Mn–Si-based shape-memory alloys (SMAs) (Sato et al. 1982; Otsuka et al. 1990). In most cases, steels contain other alloying elements, such as Cr, Ni, Al, Si, C, and N, and the concentration of Mn depends on the alloy system (e.g. 12–29 wt% in the case of binary Fe–Mn systems).

This article focuses on a design strategy for improving the tensile and fatigue properties of high-Mn steels from microstructural, thermodynamic, and crystallographic perspectives. Referring to the concept of 'plaston', a plastically deformed infinitesimal volume element, various plasticity mechanisms in high-Mn steels are visualised

T. Sawaguchi (✉)
Research Center for Structural Materials, National Institute for Materials Science, Tsukuba, Japan
e-mail: sawaguchi.takahiro@nims.go.jp

I. Tanaka et al. (eds.), *The Plaston Concept*,
https://doi.org/10.1007/978-981-16-7715-1_11

by the deformation of deltahedral (either tetrahedral or octahedral) models. These are the minimum volumetric units representing the atomic coordination in each phase. The extended dislocation glide of the γ lattice, mechanical γ twinning, and γ → ε martensitic transformation commonly involve the twinning shear of the deltahedrons as an elementary process, and the α'-martensite is associated with the Bain distortion of the deltahedrons.

Section 11.2 presents a brief review of the atomic rearrangement and the related dislocation-based models for the extended dislocation glide, mechanical γ twinning, and mechanical γ → ε martensitic transformation. In Sect. 11.3, the polyhedron models and their mutual transformations are systematically drawn for the four plasticity mechanisms, including the γ → α' martensitic transformation, along with the former three. In Sect. 11.4, the design of the tensile properties of high-Mn steels is described, which considers the selection rule of the plasticity mechanisms, microstructure development, and strain hardening mechanisms. Structures developed at the intersection of crossing habit-plane variants of the planar deformation products are highlighted. The atomic rearrangement caused by the double shears is visualised with distortion or kinking of the deltahedron models. In Sect. 11.5, the design of the fatigue properties of high-Mn steels is described. The reversible back-and-forth motion of partials with the aid of the unidirectionality of the shear deformation of the deltahedrons plays an important role in these properties.

This article addresses only microstructural, thermodynamic, and crystallographic aspects of the plasticity mechanisms. It does not address the improvement of mechanical properties by grain refinement, precipitation, and interstitial atoms, all of which have been extensively studied, especially to improve the yield stress, which is a critical weak point of TRIP/TWIP steels. For these issues, review articles (Bouaziz et al. 2011; Cooman et al. 2018, 2012; Chowdhury et al. 2017; Zambrano 2018; Lee et al. 2017; Chen et al. 2017) or other chapters of this book can be referred.

11.2 Plasticity Mechanisms in γ-austenite

The extended dislocation glide, mechanical γ twinning, and the mechanical γ → ε martensitic transformation proceed via shear displacement on the $\{1\ 1\ 1\}_\gamma$ planes, and their products commonly have a planar morphology with crystal habit on the $\{1\ 1\ 1\}_\gamma$ planes. The plated deformation products and the remaining γ-austenite form a lamellar structure. As an example of the lamellar structure, Fig. 11.1 shows a differential interference micrograph of the surface relief on an Fe–30Mn–5Si–1Al alloy, which was formed by the shear displacement of the pre-polished surface on a specimen plastically deformed to 3%. The banded surface relief is caused by the homogeneous shear associated with the mechanical γ → ε martensitic transformation, which is the predominant plasticity mechanism in this alloy. ε-Martensite bands are formed on a unique $\{1\ 1\ 1\}_\gamma$ plane in certain grains (e.g. A in Fig. 11.1), while ε-martensite plates on different $\{1\ 1\ 1\}_\gamma$ planes cross in the other grains (e.g. B in Fig. 11.1). The frequent appearance of the annealing twin boundaries, some of

Fig. 11.1 Optical
micrograph of the surface
relief on the pre-polished
surface of tensile deformed
Fe–30Mn–5Si–1Al alloy

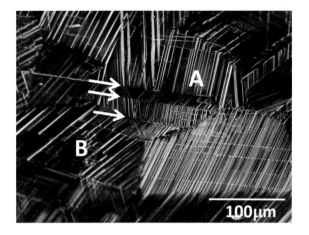

which are indicated by arrows in Fig. 11.1, is also a microstructural characteristic of
high-Mn steels with low-to-moderate SFEs.

The shear-type atomic displacements associated with the planar plasticity mechanisms on the $\{1\,1\,1\}_\gamma$ planes are shown in Fig. 11.2. With respect to the first
close-packed layer A, there are two possible second layers, B and C, with different
relative atomic positions, as shown in Fig. 11.2a, b, respectively. The atomic positions on the third $\{1\,1\,1\}_\gamma$ layer indicated in Fig. 11.2c, d) represent two types of

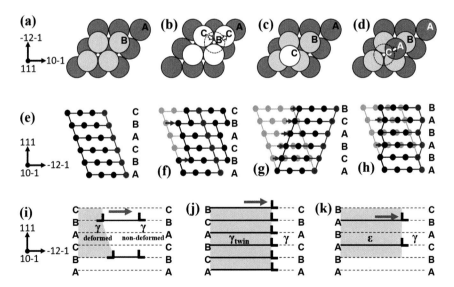

Fig. 11.2 The stacking sequence of the closest packing atomic planes and their change by the shear
displacement of the planes. **a** AB stacking unit, **b** AC stacking unit, **c** ABC stacking, **d** ABA stacking,
e γ-austenite, **f** lattice slip, **g** γ twin, **h** ε-martensite, **i** extended dislocation glide, **j** mechanical γ
twinning, and **k** mechanical γ → ε martensitic transformation

periodic stacking sequences of the close-packed layers, ABCABC or ABABAB, which correspond to the FCC and structures, respectively.

Figure 11.2e–h shows the atomic arrangements in the γ-austenite and the three plastic deformation products, which are viewed from the $[1\ 0\ 1]_\gamma$ direction. The perfect dislocation glide (e → f) does not change the stacking sequence of ABCABC, as the atoms move to the next position of the same type, for example, C to C. Mechanical twinning (e → g) is the homogeneous shear displacement of atoms, which changes the atomic positions regularly (A → B, B → A, and A → C), and the stacking sequence changes from ABCABC to ACBACB, as shown in Fig. 11.2g. The resulting structure remains unchanged, whereas the crystallographic orientation is modified into a mirror-symmetric atomic arrangement with respect to the $\{1\ 1\ 1\}_\gamma$ plane. The γ → ε martensitic transformation (e → h) is a regularised shear displacement of atoms on the alternate $\{1\ 1\ 1\}_\gamma$ planes. The stacking sequence changes from ABCABC to ABABAB, as shown in Fig. 11.2h, accompanying the structural change from FCC to HCP.

Figure 11.2i–k shows the three plasticity mechanisms as the motion of partial dislocations of the type $1/6\{1\ 1\ 1\}<1\ 1\ 2>_\gamma$ (Shockley partials). An extended dislocation in (e) consists of a few (leading and trailing) partials and stacking faults (SFs) between them. The passage of the extended dislocation causes a two-step atomic displacement from C to B and then to the next C, as shown in Fig. 11.2b, i. During this process, the stacking sequence of the SF changes locally to ABA. The mechanical γ twinning (j) and the mechanical γ → ε martensitic transformation (k) proceed via the regularised group motion of partials on every plane or on alternate $\{1\ 1\ 1\}_\gamma$ planes, respectively. In other words, they are regarded as the ordering of fully extended SFs.

Owing to the common elementary process consisting of the partial–SF units, the slip band, γ twins, and ε-martensite have similar morphological and crystallographic characteristics (plated form and crystal habit on the $\{1\ 1\ 1\}_\gamma$ plane), resulting in a generally high strain hardening capability. The selection of the plasticity mechanisms essentially depends on thermodynamic conditions, which determine the extension width of SF and the degree of ordering of SFs. However, when the plasticity mechanisms are competing thermodynamically, crystallographic orientation, and dislocation interaction can also affect the deformation microstructure.

For this reason, the Thomson tetrahedron (Fig. 11.3) is useful. The faces of the tetrahedron correspond to the $\{1\ 1\ 1\}_\gamma$ planes, and the edges of the tetrahedron correspond to the six $<1\ 1\ 0>_\gamma$ directions of the FCC structure. The corners of the tetrahedron are denoted by A, B, C, and D, and the midpoints of the opposite faces are denoted by α, β, γ, and δ. The planes opposite A, B, C, and D are denoted by a, b, c, and d, respectively, on the outer surface, and −a, −b, −c, and −d on the inner surface. The $<1\ 1\ 2>_\gamma$ Shockley partials are represented by Roman–Greek pairs (arrows from corner to mid-point). This enables us to discuss operational shears, their interactions, and the crystallographic orientation of the deformation products with respect to the crystallographic orientation of the γ-austenite, which is described in Sect. 11.4.

Fig. 11.3 Notations of the
Thomson tetrahedron

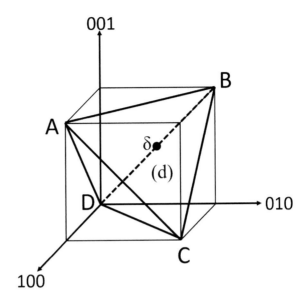

11.3 Polyhedron Models for FCC Plasticity Mechanisms

Interstitial lattice sites in FCC, HCP, and BCC structures are generally visualised by either tetrahedrons or octahedrons, whose vertices are at the centre of neighbouring atoms. These polyhedron models are widely used to discuss the distribution of interstitial atoms and their transitions during phase transformations. They can be used as unit volume elements to entirely fill the lattice spaces instead of cubic or hexagonal unit lattices. In this section, we visualise the plasticity mechanisms as distortions of the polyhedron models. Here, the extended dislocation, γ twin, and ε-martensite can be composed of a regular tetrahedron, regular octahedron, and their twin-sheared products. The α'-martensite can also be visualised by the distorted tetrahedrons produced by the so-called Bain distortion of the γ-austenite.

Figure 11.4 presents the four-unit polyhedrons. Figure 11.4a shows the atomic arrangement in the γ-austenite, in which a regular tetrahedron and regular octahedrons are drawn as the unit volume elements. Eight such tetrahedrons and four such octahedrons are included in an FCC unit lattice. The ratios between the tetrahedron and octahedron in terms of volume, number density, and space occupancy are 1:4, 2:1, and 1:2, respectively. When an a/6 (1 1 1) [$-$1 2 $-$1]$_\gamma$ partial passes on the upper atomic layer in Fig. 11.2a, the regular tetrahedron and the regular octahedron are subjected to twinning shear. Hereinafter, we refer to the distorted tetrahedron and octahedron in Fig. 11.2b as the "twin-sheared" tetrahedron and octahedron. By this twinning shear, one side of the regular tetrahedron is extended from $\left(\sqrt{2}/2\right)a_\gamma$ to a_γ, with the length of the other sides remaining unchanged, as shown in Fig. 11.2c, d. Note that the regular octahedron consists of four twin-sheared tetrahedrons, as shown

(a)

111
-12-1
10-1

(b)

(c)

$\frac{\sqrt{2}}{2}a_\gamma$

$\frac{1}{\sqrt{3}}a_\gamma$

(d)

$\frac{1}{\sqrt{3}}a_\gamma$

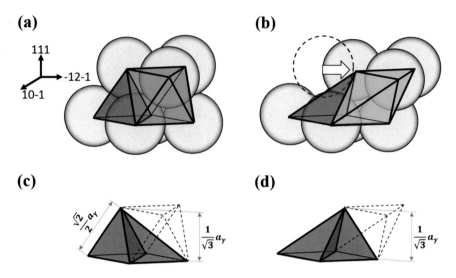

Fig. 11.4 Four-unit polyhedrons. **a** Regular tetrahedron and regular octahedron in the γ-austenite, **b** twin-sheared tetrahedron and twin-sheared octahedron after a partial passage, **c** a twin-sheared tetrahedron inside the original regular octahedron, **d** mutual transformation between the regular tetrahedron and twin-sheared tetrahedron from the **c** state

in Fig. 11.2c, while the twin-sheared octahedron consists of two regular tetrahedrons and two twin-sheared tetrahedrons, as shown in Fig. 11.2d.

Using these four-unit polyhedrons, we can fill the FCC lattice, including the SFs, γ twins, ε-martensite (= HCP lattice), and partials. An SF is a monolayer consisting of twin-sheared tetrahedrons and twin-sheared octahedrons, most of which can be replaced with regular tetrahedrons and regular octahedrons with different orientations according to the interrelation between Fig. 11.2c, d. A γ twin can be built by stacking SF monolayers, while ε-martensite can be built by alternating stacking of the SF and γ monolayers. The partial is the boundary between the twin-sheared monolayer and the γ monolayer. Along the partial, the polyhedrons are elastically distorted to adjust the atomic displacement of $a/6(1\ 1\ 1)[-1\ 2\ -1]_\gamma$.

Figure 11.5 demonstrates the deformation of the polyhedrons associated with the γ → α' martensitic transformation. The two regular tetrahedrons (a) are compressed in the 001 axis and stretched in the 100 and 010 axes (d) by the Bain distortion. The regular octahedron (b) is also distorted in the same manner (e). Hereafter, we refer to the distorted tetrahedron and octahedron in α'-martensite as the α-tetrahedron and α-octahedron, respectively. The twin-sheared tetrahedron residing in the regular octahedron (c) is transferred into the α-tetrahedron (f). Comparing Fig. 11.5e, f shows that the α-octahedron consists of four α-tetrahedrons.

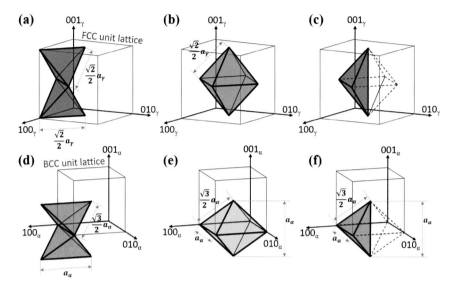

Fig. 11.5 Polyhedron models for the Bain distortion during γ → α' martensitic transformation. **a** Regular tetrahedrons in a γ unit lattice, **b** a regular octahedron in the γ unit lattice, **c** a twin-sheared tetrahedron inside the regular octahedron, **d** distorted tetrahedrons in a α' unit lattice, **e** a distorted octahedron in the α' unit lattice, **f** and a distorted tetrahedron inside the distorted octahedron

11.4   Plasticity Mechanisms Under Tensile Loading

11.4.1   Selection Rule and Generation Processes

There are several transmission electron microscopy (TEM) and high-resolution transmission electron microscopy (HRTEM) research studies on the SF as the embryo of ε-martensite and γ twins (Bray and Howe 1996; Brooks et al. 1979a, b). In designing high-Mn austenitic TRIP/TWIP steels, the stacking fault energy of the γ-austenite (SFE_γ) is used as an important indicator to predict the appearance of mechanical γ twinning and mechanical ε-martensitic transformation (Choi et al. 2020; Lee et al. 2019, 2018, 2010; Lu et al. 2017; Zambrano 2016; Das 2015; Xiong et al. 2014; Curtze et al. 2011; Nakano and Jacques 2010; Curtze and Kuokkala 2010; Tian and Zhang 2009a, b; Saeed-Akbari et al. 2009; Tian et al. 2008; Dumay et al. 2008; Cotes et al. 2004, 1999; Allain et al. 2004; Lee and Choi 2000; Grassel et al. 1997). The SFE_γ affects the extension width of the dislocation in the γ-austenite, which is determined when the repulsive force between the leading and trailing partials is equivalent to the attractive force to reduce the surface energy. The extension width increases with a decrease in the SFE_γ. The generation of the mechanical γ twins and ε-martensite requires a sufficiently wide extension width that is referred to as the 'infinite separation' (Byun 2003; Copley and Kear 1968) with respect to the grain size; thus, these plasticity mechanisms appear in low SFE_γ ranges. Remy schematically indicated the

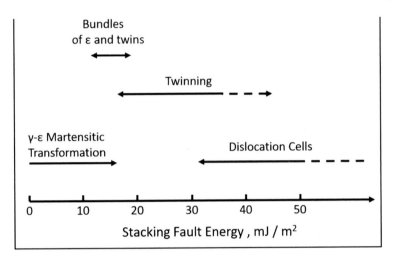

Fig. 11.6 Plasticity mechanisms observed at room temperature as a function of stacking fault energy of the austenite. (Reproduced with permission from Mater. Sci. Eng. 26 (1976) 123, Copyright 1976, Elsevier Sequoia S.A,, Lausanne)

SFE_γ ranges for the plasticity mechanisms, as shown in Fig. 11.6 (Remy and Pineau 1976). Although the threshold values differ depending on the study (Nakano and Jacques 2010; Saeed-Akbari et al. 2009; Allain et al. 2004; Pierce et al. 2014), most studies have reported that the order of appearance is dislocation glide, mechanical γ twinning, and mechanical $\gamma \rightarrow \varepsilon$ martensitic transformation with decreasing SFE_γ.

The appearance of the mechanical γ twinning and the mechanical $\gamma \rightarrow \varepsilon$ martensitic transformation in certain SFE_γ ranges is associated with their generation mechanisms. Various nucleation mechanisms have been proposed to describe how and why the SFs are regularised on every or on alternate {1 1 1} plane to form γ twins or ε-martensite, respectively. For the nucleation of the mechanical γ twin, these are listed as follows: (1) the pole mechanism (Venables 1974), (2) successive deviation of Shockley partials leaving Frank pairs at the Lomer–Cottrel lock (Cohen and Weertman 1963) (3) stair-rod cross-slip mechanism (Fujita and Mori 1975; Mori and Fujita 1980), (4) interaction of co-planar perfect dislocations into three-atomic-layer Shockley-partial embryos (Mahajan and Chin 1973), among others. Among these, (1) the pole mechanism (Hoshino et al. 1992), (3) stair-rod cross-slip mechanism (Fujita and Ueda 1972), and (4) co-planar dislocation interaction model (Mahajan et al. 1977) have also been applied to ε-martensite nucleation. Many microstructural studies have supported these models (Idrissi et al. 2010, 2013; Mahato et al. 2017; Steinmetz et al. 2013).

Microstructural observation has revealed that the overlapping of SFs into ε-martensite proceeds approximately but irregularly (Brooks et al. 1979b). For the gradual growth of the ε-martensite, the thermodynamic stability of the ε-phase is an additional requirement. The Gibbs free energy difference associated with the $\gamma \rightarrow \varepsilon$ martensitic transformation and the SFE_γ are interrelated via the following equation

(Olson and Cohen 1976):

$$\Gamma = 2\rho\left(\Delta G_{\text{chem}}^{\gamma\to\varepsilon} + \Delta G_{\text{mag}}^{\gamma\to\varepsilon}\right) + 2\sigma^{\gamma/\varepsilon}, \tag{11.1}$$

where Γ is the stacking fault energy of the austenite, ρ is the molar surface density of atoms in the (111) planes, $\Delta G_{\text{chem}}^{\gamma\to\varepsilon}$ is the chemical Gibbs free energy difference between the γ- and ε-phases, $\Delta G_{\text{mag}}^{\gamma\to\varepsilon}$ is the magnetic contribution to the Gibbs free energy differences, and $\sigma^{\gamma/\varepsilon}$ is the energy per surface unit of the (111) interface between the γ- and ε-phases. Equation (11.1) was developed based on the concept that the stacking fault is an ε-martensite embryo consisting of a double atomic layer, as shown in Fig. 11.5. There are three types of martensitic transformation paths, namely, $\gamma \to \varepsilon$, $\gamma \to \alpha$', and $\gamma \to \varepsilon \to \alpha$', depending on the Gibbs free energies of the phases, G^γ, G^ε, and G^α.

Studies have attempted to obtain reliable SFE_γ values. SFE_γ can be determined by measuring the radius of the nodes of partial dislocations using TEM (Kim and Cooman 2011; Pierce et al. 2012), measuring the extension width with the TEM weak-beam method, measuring peak broadening in XRD (Tian et al. 2008; Balogh et al. 2006; Barman et al. 2014; Schramm and Reed 1975; Tian and Zhang 2009c), and by thermodynamic and first principles calculations (Zambrano 2016; Xiong et al. 2014; Curtze et al. 2011; Nakano and Jacques 2010; Saeed-Akbari et al. 2009; Dumay et al. 2008; Allain et al. 2004; Pierce et al. 2014; Tian et al. 2017; Medvedeva et al. 2014). The critical SFE_γ values vary depending on the methods used by the investigators, alloy systems, and the method used for calculating/measuring SFE_γ. Some investigators have proposed Γ (mJ/m^2) ranges for the appearance of the plasticity mechanisms, for example, $12 < \Gamma < 35$ for mechanical γ twinning and $\Gamma < 18$ for the mechanical $\gamma \to \varepsilon$ martensitic transformation (Allain et al. 2004).

In the case of the $\gamma \to \varepsilon$ martensitic transformation, the corresponding temperature is a direct indicator for predicting its occurrence. The γ- and ε-phases have equal Gibbs free energies at the thermodynamic equilibrium temperature of T_0. The forward $\gamma \to \varepsilon$ martensitic transformation starts at $M_s < T_0$ and ends at M_f upon cooling; the reverse $\varepsilon \to \gamma$ martensitic transformation starts at $A_s > T_0$ and ends at A_f upon heating. Undercooling and overheating are required to obtain a sufficient driving force, $\Delta G^{\gamma\to\varepsilon}(M_S)$ or $\Delta G^{\gamma\to\varepsilon}(A_S)$, respectively, for the transformations. The relationship among the transformation temperatures and the driving forces is shown in Fig. 11.7a. Mechanical martensitic transformations are divided into two groups depending on the nucleation mechanism, as schematically shown in Fig. 11.7b. At temperatures immediately above M_s, spontaneous martensitic transformation occurs by a 'stress-assisted' nucleation mechanism, as the external stress reaches the critical value required to compensate the lack of a chemical driving force for the transformation. Plastic yielding occurs at the critical stress for the martensitic transformation, σ_M, that linearly increases with the deformation temperature. Above M_s^σ, σ_M becomes greater than the yielding stress by slip, σ_y. No more spontaneous transformation occurs; however, the martensitic transformation is still available below M_d

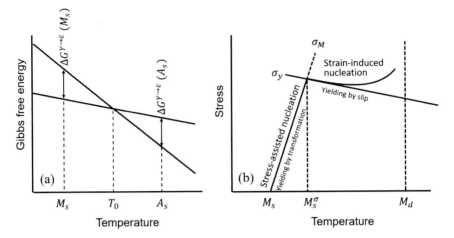

Fig. 11.7 Martensitic transformation temperatures for **a** spontaneous transformation on cooling and heating and **b** mechanical transformation with stress-assisted and strain-induced nucleation mechanisms. (Reproduced with permission from Journal of the Less Common Metals 28 (1972) 107, Copyright 1972, Elsevier Sequoia S.A,, Lausanne)

via the 'strain-induced' nucleation mechanism where dislocations act as nucleation sites for the martensite.

This concept of strain-induced nucleation was first proposed by Olson and Cohen for $\gamma \rightarrow \alpha'$ martensitic transformation (Olson and Cohen 1972). ε-Martensite was treated as a precursor or an intermediate phase during the $\gamma \rightarrow \alpha'$ martensitic transformation. The formation of α'-martensite was interpreted by a double shear mechanism, which is described in Sect. 11.4.2. Subsequently, Andersson applied the concept to the $\gamma \rightarrow \varepsilon$ martensitic transformation (Andersson and Stalmans 1998).

11.4.2 Transformation- and Twinning-Induced Plasticities

In recent decades, TWIP and TRIP have been extensively studied in high-Mn steels. The TWIP effect is interpreted to be a result of the so-called dynamic Hall–Petch effect, which is a continuous grain subdivision by planar mechanical twins during loading. This effect causes considerable strain hardening and suppresses local necking, and consequently, a good combination of high ultimate tensile stress and high elongation is available. A similar effect is expected for the mechanical $\gamma \rightarrow \varepsilon$ martensitic transformation with a similar planar morphology. Further martensitic transformation into α'-martensite can also contribute to strain hardening. In this case, however, the strain hardening is not due to the dynamic Hall–Petch effect but by the intrusion of a high density of dislocations through lattice expansion associated with the martensitic transformation. Originally, the term TRIP was given to the superior plasticity of an Fe–Ni alloy undergoing a direct $\gamma \rightarrow \alpha'$ martensitic transformation.

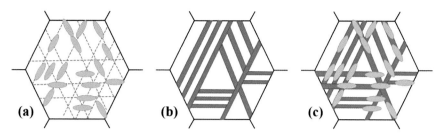

Fig. 11.8 Transformation- and twinning-induced plasticities and three types of martensitic transformation paths: **a** α-TRIP, **b** ε-TRIP and TWIP, and **c** double TRIP

The three types of TRIP effects together with the TRIP effect are summarised in Fig. 11.8.

As an example of the tensile behaviour dependent on the plasticity mechanisms, Fig. 11.9 shows the stress–strain curves of Fe–30Mn–(6-x)Si–xAl alloys (x = 0 to 6 in wt%) (Nikulin et al. 2013). In these alloys, the formation of α'-martensite is strongly inhibited, and thus, the tensile behaviour can be interpreted by ε-TRIP, TWIP, and dislocation glide. The alloy with x = 0 is a typical SMA, which plastically deforms via mechanical γ → ε martensitic transformation. As the concentration of Al that replaces Si increases, the value of SFE$_\gamma$ increases. The mechanical γ → ε martensitic transformation appears for alloys up to x = 2. The elongation increases with the change in the nucleation mechanism from stress-assisted to strain induced. The alloy with x = 2 is a typical TRIP steel, and the alloy with x = 3 is a typical TWIP steel and has the highest elongation; a further increase in the concentration of Al results in a decrease in both strength and elongation.

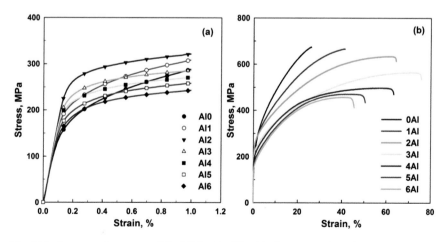

Fig. 11.9 Stress–strain curves of Fe–30Mn–(6-x)Si–xAl alloys (x = 0 to 6). (Reprinted with permission from Mater. Sci. Eng. A 587 (2013) 192, Copyright 2013, Elsevier B.V.)

The extended dislocation glide can also contribute to the strain hardening due to the suppression of the recovery, which is the stress relaxation associated with dislocation rearrangement into a cell structure. There is a variation in the dislocation structures: Yoo and Park have reported a TWIP-like hardening mechanism by a banded dislocation structure in an Fe–28Mn–9Al–0.8C alloy ($\Gamma \approx 85$ mJ/m^2), which is called microband-induced plasticity (MBIP) (Yoo and Park 2008); Gutierrez-Urrutia and Raabe have reported that multiple strain hardening is caused by dislocation substructures, such as Taylor lattices, cell blocks, and dislocation cells (Gutierrez-Urrutia and Raabe 2012).

11.4.3 Martensite/twin Variants

Martensite/twin variants are important microstructural components in high-Mn steels; these variants and γ-austenite all have crystallographic orientation relationships (ORs).

The OR of the mechanical ε-martensite with respect to the γ-austenite, which is called the Shoji–Nishiyama (S–N) OR, is shown as follows:

$\{1\ 1\ 1\}_\gamma$ // $\{0\ 0\ 0\ 1\}_\varepsilon$, $<1\ 1\ -2>_\gamma$ // $<1\ -1\ 0\ 0>_\varepsilon$. (S–N OR)

There are four ε-martensite variants depending on the habit plane denoted by a, b, c, or d, using the notation of the Thompson tetrahedron in Fig. 11.3.

A remarkable microstructural event is the formation of a new crystal phase or a new crystallographic variant by crossing shears at the intersection of different habit-plane variants of ε-martensite. In Fe–Cr–Ni stainless steels (Olson and Cohen 1972; Venables 1962; Bogers and Burgers 1964; Lee et al. 2013) and Fe–high-Mn steels, α'-martensite is reported to form at the intersection of the ε-martensite habit-plane variants. In Fe–Mn–Si SMAs, a wide variety of other intersection products have been reported, such as ε-twins (Matsumoto et al. 1994; Zhang and Sawaguchi 2018), γ-twins (Matsumoto et al. 1994), and the ε-phase rotated 90° from the original ε-phase (ε_{90}) (Yang and Wayman 1992a; b). In Fe–30Mn–4Si–2Al TRIP steel, the γ-phase rotated 90° from the parent γ-austenite (γ_{90}) (Zhang et al. 2011a; b) has also been reported.

These intersection reactions act not only as a self-accommodation mechanism but also as a source, obstacle, or sink of dislocations by providing new interfaces with surrounding phases and volumetric changes. To control the mechanical and functional properties, it is important to clarify the selection rule of the intersection reactions and their contributions to macroscopic properties. The thermodynamic stability of γ-, ε-, and α'-phases is evidently an important factor affecting the selection of the three thermodynamically distinct intersection reactions, that is, the secondary martensitic transformation into α', the reverse martensitic transformation into γ, and the plastic deformation (e.g. twinning) of ε.

Zhang et al. (2011a) pointed out that the directions of the crossing shears with respect to the intersection axis, either 90° (Type I) or 30° (Type II), also determines the intersection products. Figure 11.10 shows the EBSD images of the structures at

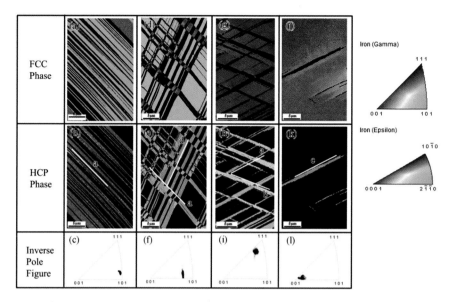

Fig. 11.10 EBSD images for structures at the intersection of ε-martensite variants in an Fe–30Mn–4Si–2Al alloy. (Reprinted with permission from Philos. Mag. Letters 91 (2011) 563, Copyright 2011, Taylor & Francis)

the intersections of crossing ε-martensite variants in Fe–30Mn–4Si–2Al steel tensile deformed to 10%. Figure 11.10a–c shows a single variant structure in the grain close to <4 1 4>$_\gamma$. Two types of double variant structures are shown at an orientation on the <0 0 1> − <1 0 1>$_\gamma$ boundary in Fig. 11.10d–f and the <0 0 1> − <1 1 1> γ boundary in Fig. 11.10g–i. Using Schmid's law, plausible crossing shears are identified as Type I for the former and Type II for the latter. The γ 90° phase was observed at the Type I intersection, while a non-basal slip was observed at the Type II intersection. At an orientation close to <0 0 1>$_\gamma$ in Fig. 11.10j–l, the martensite fraction is significantly lower than that of the other grains because of the low Schmid factors of Shockley partials compared with those of perfect dislocations.

The orientation dependence and double shear mechanisms for various intersection products are summarised in Fig. 11.11. Starting from the γ-austenite (a) represented by the regular tetrahedron, (b) one-third, **T/3**, (c) half, **T/2**, and (d) full twinning, **T**, shears along Cδ/d are considered as intermediate states. Bogers and Burgers proposed that the second shear on the non-conjugate plane with a magnitude of **T/2** can produce α'-martensite (Bogers and Burgers 1964). Figure 11.10e is one such example produced by the second shear along Bα/a. This double shear mechanism is expected at the Type II (30°) intersection. From the half-twin state (c), ε-martensite (f) is produced by shuffling on the d-plane, γ$_{90}$ is produced by T/2 shear on the c-plane, and {1 0 −1 2}ε twinning is produced by shuffling on the c-plane. The latter two are both Type I intersection products. The double shear mechanism for the γ$_{90}$ phase was first schematically drawn by Sleeswyk (Sleeswyk 1962) and

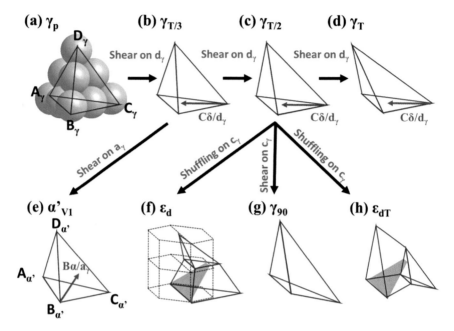

Fig. 11.11 Double shear mechanisms for various intersection products

experimental observation of the γ_{90} and the $\{1\,0-1\,2\}\varepsilon$ twin at Type I intersections was confirmed by EBSD and TEM in Zhang et al. 2011b.

11.5 Plasticity Mechanisms Under Cyclic Loading

Remy and Chalant reported a positive effect of the planar plasticity mechanisms on fatigue life (Chalant and Remy 1980). Suppression of the cross slip of dislocations owing to planarity is considered to decelerate the accumulation of fatigue damage. Figure 11.12 schematically shows the reversible back-and-forth movement of the partials associated with martensitic transformations, twinning, and dislocation glide (Sawaguchi et al. 2015). These mechanisms commonly involve the movement of a partial dislocation connected with an SF as an elementary step. All of these are expected to improve the fatigue life; however, the degree of improvement is different, probably due to different degrees of reversibility.

Recently, studies have extensively investigated the reversal of mechanical twinning and its positive effect on fatigue properties. Direct observations of the twinning and detwinning in TWIP steels under cyclic loading have been made using an in situ neutron diffraction technique (Xie et al. 2018) and by EBSD analysis on interrupted reverse loading (McCormack et al. 2018).

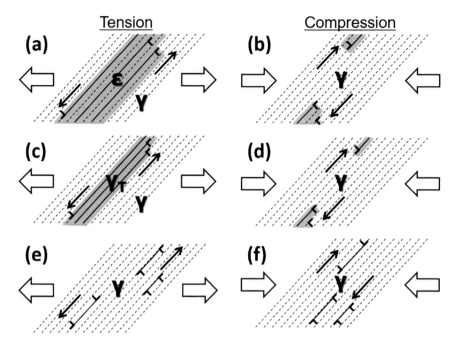

Fig. 11.12 Reversible movement of Shockley-partial dislocations associated with **a, b** mechanical $\gamma \rightarrow \varepsilon$ martensitic transformation, **c, d** mechanical γ twinning, and **e, f** extended dislocation glide. (Reprinted with permission from Scripta Mater. 99 (2015) 49, Copyright 2014, Acta Materialia Inc. Published by Elsevier Ltd.)

Sawaguchi et al. reported a reversible change in the surface relief formed on the pre-polished surface of an Fe–28Mn–6Si–5Cr–0.5NbC SMA during tensile and subsequent compressive loadings (Sawaguchi et al. 2006). They suggested possible improvement of the fatigue resistance by the reversible deformation characteristics of the SMA. However, subsequent investigations revealed that the optimum chemical composition was different from that of the SMA.

Figure 11.13 shows the fatigue lives of Fe–30Mn–(6-x)Si–xAl alloys with different compositions (x = 0–6) at room temperature after strain-controlled tension–compression loading at a total strain amplitude of 0.01 (Nikulin et al. 2013). Note that the alloys are those discussed in Sect. 11.4.2. regarding their tensile behaviour. The Fe–30Mn–4Si–2Al TRIP steel exhibits the longest value of the fatigue life (N_f), whereas the Fe–30Mn–6Si SMA and the Fe–30Mn–3Si–3Al TWIP steels possess lower values of N_f than those of the former. SMAs can also exhibit improved Nf when the deformation temperature is increased. Figure 11.14 represents the fatigue lives of an Fe–28Mn–6Si–5Cr SMA after strain-controlled tension–compression loading at a total strain amplitude of 0.01 at temperatures ranging from 223 to 474 K.

The composition and temperature dependence of the fatigue life of high-Mn steels reached a maximum when they were deformed in the vicinity of the M_d temperature. The temperature range is characterised by the strain-induced nucleation of martensite,

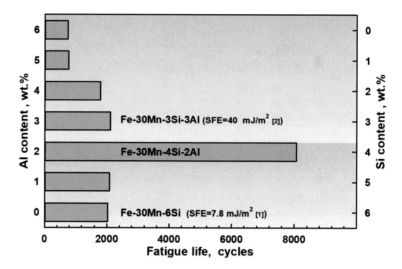

Fig. 11.13 Fatigue lives of Fe–30Mn–(6-x)Si–xAl alloys (x = 0 to 6) for cyclic tension–compression loading at a strain amplitude of 0.01. (Reproduced with permission from Mater. Sci. Eng. A 587 (2013) 192, Copyright 2013, Elsevier B.V.)

as discussed in Sect. 11.4.1. The accumulated volume fraction of the ε-martensite increases slowly under cyclic loading, and the post-fatigue fractured microstructure shows a dual γ/ε phase. An example of such dual phase microstructure observed on the Fe–30Mn–4Si–2Al TRIP steel by SEM-EBSD is shown in Fig. 11.15b, while the single ε-phase microstructure on the Fe–30Mn–6Si SMA and the single γ-phase microstructure on the Fe–30Mn–3Si–3Al TWIP steels are shown in Fig. 11.15a, c, respectively.

A possible reason for the highest N_f value occurring at this temperature is the small difference in the Gibbs free energy between the γ- and ε-phases. The Shockley-partial shear occurs only in such a way that the atom goes through the saddle point between the two atoms on the lower layer, as shown in Fig. 11.4, while the movement in the opposite direction is strongly forbidden. After partial shear, however, reversal shear can easily occur under counter-directional loading. Owing to the unidirectionality, the Shockley-partial shear yields reversibility under cyclic loading. However, if the Gibbs energy of ε-martensite is sufficiently low, once ε-martensite is formed, it cannot reversely transform into austenite. That is, the thermodynamic condition of $\Delta G \lesssim 0$ is required for the reversible martensitic transformation. Such reversible twinning shear is also expected in the twinning–detwinning process. So far, however, the N_f values at temperatures above M_d are lower than those at temperatures below M_d. This is probably due to complicated dislocation interactions at relatively higher SFE_γ values with narrower dislocation extension widths.

A new application of high-Mn steels with improved fatigue life has been proposed for the seismic response control of architectural constructions. Steel seismic dampers have recently been used to protect buildings by elasto-plastic deformation hysteresis

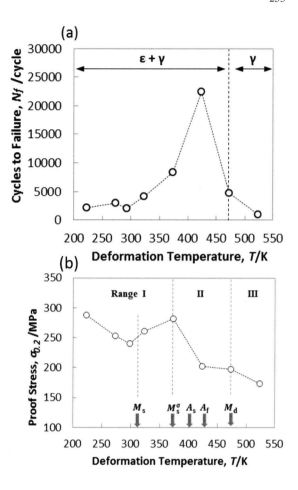

Fig. 11.14 Fatigue life **a** and 0.2% proof stress **b** as a function of deformation temperature with respect to martensitic transformation temperatures (Tasaki et al. 2016). (Reprinted with permission from Mater. Trans. 57 (2016) 639, Copyright 2016, Japan Institute of Metals and Materials)

that transforms seismic energy into heat. A new seismic damping alloy, Fe–15Mn–10Cr–8Ni–4Si, was developed based on the design concept described in this article and is currently used in some large-scale buildings (Sawaguchi et al. 2016).

11.6 Concluding Remarks

An attempt is made to describe the plasticity mechanisms in high-Mn austenitic steels in a relatively comprehensive manner, considering the distortions of poly-hedron models. The selection rule of the plasticity mechanisms, microstructural development, and their effects on the tensile and fatigue properties are discussed in terms of microstructural, thermodynamic, and crystallographic aspects. The tensile and fatigue properties of coarse-grained and interstitial-free Fe–Mn–Si–Al alloys

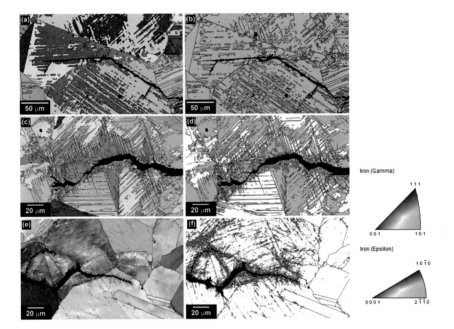

Fig. 11.15 EBSD-IPF maps **a, c,** and **e** and phase/boundary maps **b, d,** and **f** of the fatigue fractured specimens of **a, b** Fe–30Mn–6Si, **c, d** Fe–30Mn–4Si–2Al, and **e, f** Fe–30Mn–3Si–3Al. (Reprinted with permission from Mater. Sci. Eng. A 587 (2013) 192, Copyright 2013, Elsevier B.V.)

with different SFE_γ values are shown as representatives. The effects of grain refinement, precipitation, and interstitial elements are excluded here. Nevertheless, some microstructural events, such as crossing shear reactions at the intersection of the different ε-martensite variants and reversible back-and-forth movement of Shockley, which are highlighted in this article, are of importance in designing mechanical and fatigue properties of high-Mn steels.

References

Allain S, Chateau JP, Bouaziz O, Migot S, Guelton N (2004) Mater Sci Eng A-Struct Mater Prop Microstruct Process 387:158–162
Andersson M, Stalmans R, ÅgrenJ (1998) Acta Materialia 46(11):3883–3891
Balogh L, Ribárik G, Ungár T (2006) J Apple Phys 100(2):023512
Barman H, Hamada AS, Sahu T, Mahato B, Talonen J, Shee SK, Sahu P, Porter DA, Karjalainen LP (2014) Metall Mater Trans A 45(4):1937–1952
Bogers AJ, Burgers WG (1964) Acta Metall 12(2):255–261
Bouaziz O, Allain S, Scott CP, Cugy P, Barbier D (2011) Curr Opin Sol Stat Mater Sci 15(4):141–168
Bray DW, Howe JM (1996) Metall Mater Trans A 27A(11):3371–3380
Brooks J, Loretto M, Smallman R (1979a) Acta Metall 27(12):1839–1847
Brooks J, Loretto M, Smallman R (1979b) Acta Metall 27(12):1829–1838

Byun TS (2003) Acta Mater 51(11):3063–3071
Chalant G, Remy L (1980) Acta Metall 28(1):75–88
Charles J, Berghezan A, Lutts A, Dancoisne PL (1981) Metal Progress 119(6):71–74
Chen S, Rana R, Haldar A, Ray RK (2017) Prog Mater Sci 89:345–391
Choi YW, Dong ZH, Li W, Schonecker S, Kim H, Kwon SK, Vitos L (2020) Mater Des 187:8
Chowdhury P, Canadinc D, Sehitoglu H (2017) Mater Sci Eng: r: Rep 122:1–28
Cohen JB, Weertman J (1963) Acta Metall 11(8):996–998
Copley SM, Kear BH (1968) Acta Metall 16(2):227–231
Cotes S, Guillermet AF, Sade M (1999) Mater Sci Eng A-Struct Mater Prop Microstruct Process
    273:503–506
Cotes SM, Guillermet AF, Sade M (2004) Metall Mater Trans A-Phys Metall Mater Sci 35A(1):83–
    91
Curtze S, Kuokkala VT (2010) Acta Mater 58(15):5129–5141
Curtze S, Kuokkala VT, Oikari A, Talonen J, Hanninen H (2011) Acta Mater 59(3):1068–1076
Das A (2015) Metall Mater Trans A 47(2):748–768
De Cooman BC, Kwon O, Chin KG (2012) Mater Sci Technol 28(5):513–527
De Cooman BC, Estrin Y, Kim SK (2018) Acta Mater 142:283–362
Dumay A, Chateau JP, Allain S, Migot S, Bouaziz O (2008) Mater Sci Eng A-Struct Mater Prop
    Microstruct Process 483–84:184–187
Fujita H, Mori T (1975) Scripta Metall 9(6):631–636
Fujita H, Ueda S (1972) Acta Metall 20(5):759–767
Grassel O, Frommeyer G, Derder C, Hofmann H (1997) J Phys IV 7(C5):383–388
Grassel O, Kruger L, Frommeyer G, Meyer LW (2000) Int J Plast 16(10–11):1391–1409
Gutierrez-Urrutia I, Raabe D (2012) Acta Mater 60(16):5791–5802
Hadfield RA (1888) Science 284–286
Hoshino Y, Nakamura S, Ishikawa N, Yamaji Y, Matsumoto S, Tanaka Y, Sato A (1992) Mater
    Trans JIM 33(3):253–262
Idrissi H, Renard K, Ryelandt L, Schryvers D, Jacques PJ (2010) Acta Mater 58(7):2464–2476
Idrissi H, Renard K, Schryvers D, Jacques PJ (2013) Philos Mag 93(35):4378–4391
Jee KK, Jang WY, Baik SH, Shin MC, Choi CS (1997) Scripta Mater 37(7):943–948
Kim J, De Cooman BC (2011) Metall Mater Trans A 42(4):932–936
Kim H, Ha Y, Kwon KH, Kang M, Kim NJ, Lee S (2015) Acta Mater 87:332–343
Lee YK, Choi CS (2000) Metall Mater Trans A-Phys Metall Mater Sci 31(2):355–360
Lee TH, Shin E, Oh CS, Ha HY, Kim SJ (2010) Acta Mater 58(8):3173–3186
Lee T-H, Ha H-Y, Kang J-Y, Moon J, Lee C-H, Park S-J (2013) Acta Mater 61(19):7399–7410
Lee SJ, Han J, Lee S, Kang SH, Lee SM, Lee YK (2017) Sci Rep 7(1):3573
Lee S-J, Fujii H, Ushioda K (2018) J Alloy Compd 749:776–782
Lee S-J, Ushioda K, Fujii H (2019) Mater Charact 147:379–383
Lu S, Li RH, Kadas K, Zhang HL, Tian YZ, Kwon SK, Kokko K, Hu QM, Hertzman S, Vitos L
    (2017) Acta Mater 122:72–81
Mahajan S, Green M, Brasen D (1977) Metall Mater Trans A 8(2):283–293
Mahajan S, Chin GY (1973) 5th Spring Meeting of the Metallurgical Society of AIME (abstracts
    only received)|5th Spring Meeting of the Metallurgical Society of AIME (abstracts only received)
    47|ii+155
Mahato B, Sahu T, Shee SK, Sahu P, Sawaguchi T, Komi J, Karjalainen LP (2017) Acta Mater
    132:264–275
Matsumoto S, Sato A, Mori T (1994) Acta Metall Mater 42(4):1207–1213
McCormack SJ, Wen W, Pereloma EV, Tome CN, Gazder AA, Saleh AA (2018) Acta Mater
    156:172–182
Medvedeva N, Park M, Van Aken DC, Medvedeva JE (2014) J Alloy Compd 582:475–482
Mori T, Fujita H (1980) Acta Metall 28(6):771–776
Nakano J, Jacques PJ (2010) Calphad 34(2):167–175

Nikulin I, Sawaguchi T, Tsuzaki K (2013) Mater Sci Eng A-Struct Mater Prop Microstruct Process 587:192–200
Olson GB, Cohen M (1972) J Less Common Metals 28(1):107–118
Olson GB, Cohen M (1976) Metall Trans A 7(12):1897–1904
Otsuka H, Yamada H, Maruyama T, Tanahashi H, Matsuda S, Murakami M (1990) ISIJ Int 30(8):674–679
Pierce DT, Bentley J, Jimenez JA, Wittig JE (2012) Scripta Mater 66(10):753–756
Pierce DT, Jimenez JA, Bentley J, Raabe D, Oskay C, Wittig JE (2014) Acta Mater 68(15):238–253
Remy L, Pineau A (1976) Mater Sci Eng 26(1):123–132
Saeed-Akbari A, Imlau J, Prahl U, Bleck W (2009) Metall Mater Trans A-Phys Metall Mater Sci 40A(13):3076–3090
Sato A, Chishima E, Soma K, Mori T (1982) Acta Metall 30(6):1177–1183
Sawaguchi T, Kikuchi T, Ogawa K, Kajiwara S, Ikeo Y, Kojima M, Ogawa T (2006) Mater Trans 47(3):580–583
Sawaguchi T, Nikulin I, Ogawa K, Sekido K, Takamori S, Maruyama T, Chiba Y, Kushibe A, Inoue Y, Tsuzaki K (2015) Scripta Mater 99:49–52
Sawaguchi T, Maruyama T, Otsuka H, Kushibe A, Inoue Y, Tsuzaki K (2016) Mater Trans 57(3):283–293
Schramm R, Reed R (1975) MTA 6(7):1345
Shin S, Kwon M, Cho W, Suh IS, De Cooman BC (2017) Mater Sci Eng A-Struct Mater Prop Microstruct Process 683:187–194
Sleeswyk AW (1962) Philos Mag 7(81):1597
Sohn SS, Hong S, Lee J, Suh BC, Kim SK, Lee BJ, Kim NJ, Lee S (2015) Acta Mater 100:39–52
Steinmetz DR, Japel T, Wietbrock B, Eisenlohr P, Gutierrez-Urrutia I, Saeed-Akbari A, Hickel T, Roters F, Raabe D (2013) Acta Mater 61(2):494–510
Tasaki W, Sawaguchi T, Nikulin I, Sekido K, Tsuchiya K (2016) Mater Trans 57(5):639–646
Tian X, Zhang YS (2009a) Mater Sci Eng A-Struct Mater Prop Microstruct Process 516(1–2):78–83
Tian X, Zhang YS (2009c) Mater Sci Eng A-Struct Mater Prop Microstruct Process 516(1–2):73–77
Tian X, Li H, Zhang Y (2008) J Mater Sci 43(18):6214–6222
Tian L-Y, Lizarraga R, Larsson H, Holmstrom E, Vitos L (2017) Acta Mater 136:215–223
Tian X, Zhang Y (2009) Mater Sci Eng A 516(1–2):73–77
Venables JA (1962) Philos Mag 7(73):35–000
Venables JA (1974) Philos Mag A 30:1165–1169
Wang HJ, Wang H, Zhang RQ, Liu R, Xu Y, Tang R (2019) J Alloy Compd 770:252–256
Xie Q, Chen Y, Yang P, Zhao Z, Wang YD, An K (2018) Scripta Mater 150:168–172
Xiong RL, Peng HB, Si HT, Zhang WH, Wen YH (2014) Mater Sci Eng A-Struct Mater Prop Microstruct Process 598:376–386
Yang JH, Wayman CM (1992a) Acta Metall Mater 40(8):2011–2023
Yang JH, Wayman CM (1992b) Acta Metall Mater 40(8):2025–2031
Yoo JD, Park KT (2008) Mater Sci Eng A-Struct Mater Prop Microstruct Process 496(1–2):417–424
Zambrano OA (2016) J Eng Mater Technol-Trans ASME 138(4):9
Zambrano OA (2018) J Mater Sci 53(20):14003–14062
Zhang X, Sawaguchi T, Ogawa K, Yin F, Zhao X (2011a) Philos Mag Lett 91(9):563–571
Zhang X, Sawaguchi T, Ogawa K, Yin F, Zhao X (2011b) Philos Mag 91(35):4410–4426
Zhang X, Sawaguchi T (2018) Acta Materialia 143(Supplement C):237–247

Chapter 12
Design and Development of Novel Wrought Magnesium Alloys

Taisuke Sasaki and Kazuhiro Hono

12.1 Introduction

Magnesium alloy has received revived research interest owing to ever increasing demand for lightweight materials in the transportation sector to lower the emission of greenhouse effect gases by weight reduction. However, there are only a few applications of magnesium alloys such as cast alloys for engine blocks. If wrought magnesium alloys can substitute some of the structural components in automotive bodies, further reduction of weight is expected. However, there are limited applications of wrought alloys due to their poor mechanical properties, low corrosion resistance, flammability, and high cost of processing. Therefore, the development of novel wrought alloys is essential to broaden the applications of magnesium alloys to automotive bodies. This chapter discusses the requirements to make wrought magnesium alloys industrially viable through a brief review of their mechanical properties. A new concept to achieve improved mechanical properties is presented along with examples of novel heat treatable wrought magnesium alloys.

T. Sasaki (✉) · K. Hono
National Institute for Materials Science (NIMS), Tsukuba 305-0044, Ibaraki, Japan
e-mail: sasaki.taisuke@nims.go.jp

K. Hono
e-mail: kazuhiro.hono@nims.go.jp

T. Sasaki
Center for Elements Strategy Initiative for Structural Materials (ESISM), Kyoto University, Sakyo-ku, Kyoto 606-8501, Japan

12.2 Requirements for Wrought Magnesium Alloys

The majority of structural components in transportation vehicles consist of steel and aluminum alloys. If wrought magnesium alloys are to substitute some of these materials, the best target materials would be Al–Mg–Si-based (6XXX series) alloys, which are used as body panels, door frames, and space frames.

Certain magnesium wrought alloys have equivalent or superior mechanical properties compared with those of 6XXX series aluminum alloys as shown in Fig. 12.1 (b). However, applications of wrought magnesium alloys are limited. One main reason for this is the high processing cost that originates from their poor workability and formability. Low corrosion susceptibility and flammability would also limit the applications. However, the poor flammability can be improved by the addition of Ca. The following sections discuss the directions for developing industrially viable wrought magnesium alloys emphasizing the reduction of processing costs.

12.2.1 Extruded Alloys

Figure 12.2 shows an extrusion limit window diagram for various magnesium alloys and AA6063 aluminum alloy (Atwell and Barnett, 2007). This figure shows extrusion temperature ranges and speeds for successful extrusion. Successful extrusion is expected if the extrusion temperature and speed are within the process windows.

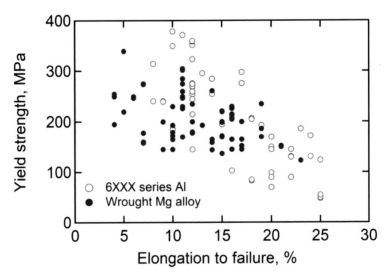

Fig. 12.1 Relationship between strength and elongation to failure in 6XXX series aluminum alloys and wrought magnesium alloys. Data points are adopted from Davis (1993); Abedesian and Baker (1999); Kamado et al. (2005)

Fig. 12.2 Extrusion limit window diagram for Mg–xAl–1Zn (x = 3, 6) (wt.%) alloys. Reprinted from ref. Atwell and Barnett (2007) with permission from Springer Nature

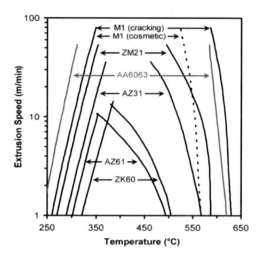

Mg–6Al–1Zn (AZ61) and Mg–6Zn–0.5Zr (ZK60) are commercial wrought magnesium alloys with yield strengths of ~ 230 MPa, which is comparable with that of the AA6063 alloy. However, AZ61 and ZK60 alloys have poor extrudability compared with AA6063. This means that the processing cost for AZ61 and ZK60 is much higher than that for AA6063. Therefore, the key to make these magnesium alloys industrially viable is to substantially improve their extrudability.

Figure 12.2 also shows that better extrudability may be achieved by lowering the composition of alloying elements; the extrusion limit windows for Mg–3Al–1Zn (AZ31), Mg–2Zn–1Mn (ZM21), and Mg–1Mn (M1) alloys are much larger than those for the AZ61 and ZK60 alloys. However, lowering the content of alloying elements is accompanied by a reduction in the yield strengths (Murai, 2004). Murai et al. reported that the maximum extrusion speed can be increased by lowering the Al content in Mg–xAl alloys from x = 2.0 to 0.5. However, the decrease in Al content results in the decrease of the yield strength from 180 to 140 MPa, respectively. The use of precipitation hardenable "dilute" alloy might be a promising approach towards the achievement of an excellent extrudability and high strength simultaneously because the dilute alloy is solution treatable and expected to be extrudable in the single-phase region. The extruded products are expected to be strengthened by subsequent artificial aging. This is the design concept for high-strength magnesium alloys with excellent extrudability.

12.2.2 Sheet Alloys

Room temperature formability is one of the important properties required for sheet alloys as a low processing cost is anticipated. Although aluminum alloy and steel sheets are easily processed into final shapes by stamping or other forming

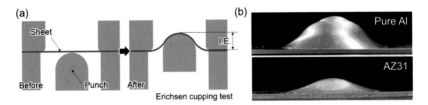

Fig. 12.3 (**a**) Schematic illustration showing the principle of the Erichsen cupping test. (**b**) Snapshots of Mg–3Al–1Zn (wt.%) (AZ31) alloy and 6XXX series aluminum alloy sheets after the Erichsen cupping test

processes at room temperature, magnesium alloy sheets need elevated temperature environments of ~300 °C for the forming process due to the poor formability at room temperature. The Erichsen cupping test is widely used as an indicator of the formability of sheet alloys. It evaluates the stretch formability in Index Erichsen (I.E.) value, which corresponds to the height required to fracture a cupped sheet metal deformed by a spherically shaped punch (Fig. 12.3 (a)). Figure 12.3 (b) shows snapshots of the samples after the Erichsen cupping test, demonstrating that the AZ31 commercial alloy sheet has a much lower stretch formability than a pure aluminum sheet; the I.E. value for the AZ31 sheet is only 2 ~ 3 mm, which is much lower than that of pure aluminum alloy (8.9 mm).

The poor room temperature formability of the magnesium alloy sheet is attributed to the limited number of slip systems due to the hexagonal close-packed (hcp) structure and the strong (0002) texture that develops during the hot-rolling process. As schematically shown in Fig. 12.4 (a), the [0002] directions (c-axes) of the magnesium grains are strongly aligned to the sheet normal direction (ND) during the hot-rolling process. In such a case, the deformation along the sheet thickness direction is difficult because the slip system is limited to the basal slip at room temperature in magnesium alloys (Fig. 12.4 (b)).

Weakening the basal texture is known to be an effective approach toward improving the poor room temperature formability. For example, the strong (0002) texture is weakened by the addition of trace Ce into a Mg–1Zn (wt.%) alloy (Mackenzie and Pekguleryuz, 2008). Chino et al. showed that the Mg–1Zn–0.2Ce

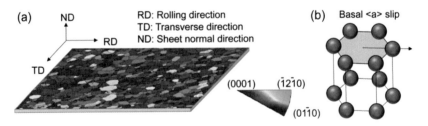

Fig. 12.4 Schematic illustration of (**a**) a typical crystallographic texture developed in magnesium alloy sheets and (**b**) illustration of basal < a > slip in magnesium and magnesium alloys

(wt.%) (ZE10) alloy exhibits a high I.E. value of 8–9 mm, which was comparable with that of Al–Mg–Si (6XXX series) aluminum alloys (Chino et al. 2008, 2010a). The development of such weakened texture was also reported following the addition of Ca or rare-earth (RE) elements into Mg–Zn-based alloys (Chino et al. 2009, 2010b, 2011; Wu et al. 2011; Bhattacharjee et al. 2014; Bian et al. 2016; Park et al. 2017), and the optimization of the thermo-mechanical processing route in Mg–Zn and Mg–Al alloys (Chino and Mabuchi 2009; Huang et al. 2009, 2011, 2012, 2015a, b; Suzuki et al. 2015; Yi et al. 2016; Trang et al. 2018).

Figure 12.5 shows the relationship between the yield strengths,σ_{ys} and I.E. values for various magnesium alloys. Some magnesium alloys have comparable stretch formability with that of aluminum alloys. However, there is a trade-off relationship between the I.E. value and the yield strength, σ_{ys}, that is, high I.E. values are achieved at the expense ofσ_{ys}. Therefore, because a value of σ_{ys} below 150 MPa is not satisfactory for automobile body applications, another approach is required to achieve excellent room temperature formability and satisfactory strength simultaneously.

To overcome the tread-off relationship between yield strength and stretch formability, the use of a precipitation hardenable alloy is regarded as a promising approach. As shown in Fig. 12.6, a solution treatment is applied to the hot-rolled sheet to form a supersaturated solid solution and recrystallized microstructure. If the recrystallized microstructure shows a weak (0001) texture, the solution-treated sheet may yield a good room temperature formability. The low strength of the solution-treated sample

Fig. 12.5 Relationship between the yield strengths, σ_{ys} and I.E. values for various magnesium alloys. The figure also includes the data for 6XXX series aluminum alloys

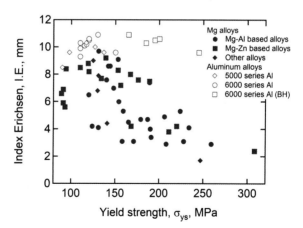

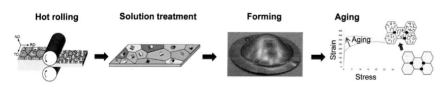

Fig. 12.6 Illustration of the processing route of heat-treatable alloys

can be substantially increased by subsequent artificial aging. This is a new design concept for developing formable high-strength magnesium sheet alloys.

12.3 Development of Industrially Viable Precipitation Hardenable Alloys

As most binary magnesium alloys are based on systems with potential for precipitation hardening, selection of the alloys that exhibit precipitation hardening is relatively simple. However, most precipitation hardenable alloys reported in the literature are not industrially viable because of their sluggish precipitation kinetics and poor age hardenability. Figure 12.7 (a) shows the variation in Vickers hardness (VHN) as a function of aging time for various binary magnesium alloys. For comparison, the age-hardening curves for Al–Cu–Sn alloy are also shown (Ringer et al. 1995; Nie et al. 2005; Sasaki et al. 2006; Mendis et al. 2007; Oh-ishi et al. 2009). Binary magnesium alloys show very sluggish kinetics for age hardening and demonstrate a relatively small hardness increment by artificial aging compared to Al–Cu–Sn alloys. Most magnesium alloys take tens of hours to reach their peak hardness through artificial aging. Thus, for making the precipitation hardening usable in heat-treatable wrought alloys, the kinetics for precipitation hardening must be substantially accelerated.

In most of the binary magnesium alloys, precipitates form on the (0002) plane (basal plane) of the magnesium matrix. Since the basal slip is the main deformation

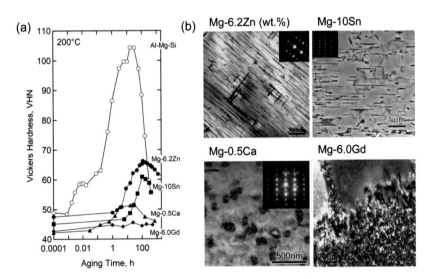

Fig. 12.7 (a) Variations in Vickers hardness as functions of aging time for various binary magnesium alloys and Al–Cu alloy. (b) Bright-field TEM images obtained from various magnesium alloys at their peak aged conditions. Reprints from ref. Ringer et al. (1995); Nie et al. (2005); Sasaki et al. (2006); Mendis et al. (2007); Oh-ishi et al. (2009), Copyright 2012, with permission from Elsevier

mechanism in magnesium alloys, the dispersion of the fine plate-like precipitates forming on the prismatic planes of the matrix is expected to be the most effective for precipitation strengthening (Nie, 2003). However, the precipitation of the "prismatic plate" is limited to magnesium–rare earth elements (Mg–RE) systems (Nie, 2012). Since the precipitates observed in the peak-aged binary alloys are coarse and heterogeneously dispersed (Fig. 12.7 (b)), a practical approach to achieve the high peak hardness would be to refine the precipitates via micro-alloying, double aging, or stretch aging. Recent studies on precipitation hardenable magnesium alloys showed that micro-alloying is effective to increase the precipitation hardening response substantially, and very high peak hardness values of ~ 100 VHN were achieved in Mg–Sn and Mg–Zn-based alloys (Fig. 12.8) (Mendis et al. 2007; Elsayed et al. 2013; Sasaki et al. 2015).

However, excellent extrudability or room temperature formability may not be anticipated from these alloys as they contain a high content of alloying elements. In search of precipitation hardenable RE-free alloys, Oh-ishi et al. reported that the age-hardening response of dilute Mg–Ca alloys can be substantially enhanced by the trace addition of Zn and Al because the strengthening phase changes from the equilibrium Mg_2Ca phase, which precipitates heterogeneously with an incoherent interface, to fully coherent Guinier Preston (G.P.) zones (Oh-ishi et al. 2009). The formation of the G.P. zone in the Mg–Ca system was first reported by Oh et al. (Oh et al. 2005). They reported that the trace addition of Zn in the Mg–0.5Ca (wt.%) alloy resulted in the formation of the monolayer G.P. zone enriched with Ca and Zn. Later, Oh-ishi et al. thoroughly investigated the effect of the Zn content on the age hardening response in the Mg–Ca–Zn alloy (Oh-ishi et al. 2009), reporting that the highest age-hardening response is achieved by the addition of 1.6 wt.% of Zn, and the addition of excess Zn results in the change of the precipitate phase. Oh-ishi et al. also identified the structure of the monolayer G.P. zones using atomic resolution high angle annular dark-field scanning transmission electron microscope (HAADF-STEM) observations, as presented in Fig. 12.9 (a). In the G.P. zone, Ca or Zn occupies the nearest-neighbor Mg site at intervals of two Mg atoms, resulting in the ordered structure as shown in Fig. 12.9 (b).

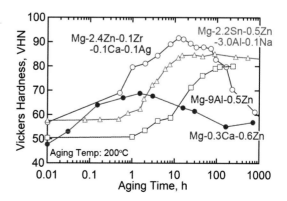

Fig. 12.8 Variations in Vickers hardness as functions of aging time for various magnesium alloys. Data points adopted from ref. Mendis et al. (2007); Oh-ishi et al. (2009); Elsayed et al. (2013)

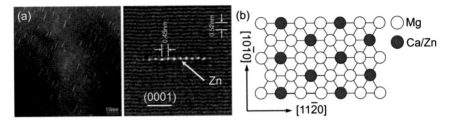

Fig. 12.9 HAADF-STEM image obtained from the peak aged Mg–0.5Ca–1.6Zn alloy (**a**) and Schematics of the ordered G.P. zone formed on a single basal plane of the Mg (**b**). Reprinted from ref. Oh-ishi et al. (2009), Copyright 2012, with permission from Elsevier

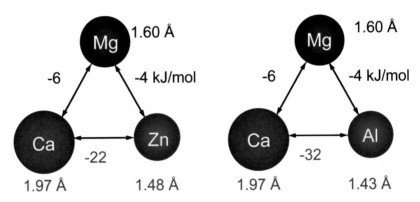

Fig. 12.10 Relationship between atomic radii and enthalpy of mixing among alloying elements and magnesium in Mg–Ca–Zn and Mg–Ca–Al alloys

Figure 12.10 shows the relationship between atomic radii and the enthalpy of mixing among Mg, Ca, and Zn. These elements have a negative enthalpy of mixing, and the atomic radius of Ca is larger than that of Mg, and that of Zn is smaller than that of Mg. This leads to the internal order of atoms within the monoatomic layer of the basal plane as shown in Fig. 12.9. In addition to Zn, Al also has a large negative enthalpy of mixing with Ca with a smaller atomic radius. Thus, Al was also expected to work to induce the precipitation of ordered G.P. zones.

Jayajaj et al. proved that the G.P. zone also forms by adding a trace amount of Al into the Mg–Ca system. As shown by the variation in the Vickers hardness as a function of the aging time (Fig. 12.11), micro-alloying Al into the Mg–0.5Ca alloy results in a substantial increase in the peak hardness along with a kinetics acceleration (Jayaraj et al. 2010). While the Mg–0.5Ca alloy shows only a small hardness increment, the Mg–0.5Ca–0.3Al alloy shows the higher peak hardness of 72 VHN in only 2 h of aging at 200 °C. As shown in a bright-field (BF) TEM image of the peak-aged Mg–0.5Ca–0.3Al alloy (Fig. 12.11 (b)), the ordered G.P. zone is precipitated as evident from the diffraction contrasts and the streaking feature in

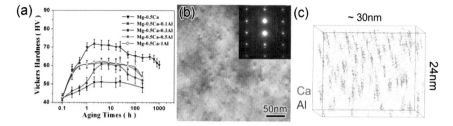

Fig. 12.11 (**a**) Variations in Vickers hardness as functions of time for Mg–0.5Ca–xAl (x = 0.1, 0.3, 0.5, and 1.0) alloys during artificial aging at 200 °C, (**b**) Mg–0.5Ca–0.3Al alloys, (**c**) 3D atom map of Al and Ca taken from the Mg–0.5Ca–0.3Al alloy. Reprinted from ref. Jayaraj et al. (2010), Copyright 2012, with permission from Elsevier

the selected area diffraction (SAD) pattern. A 3D atom map of the peak-aged Mg–0.5Ca–0.3Al alloy shows that Al and Ca are enriched in the G.P. zones (Fig. 12.11 (c)).

12.4 Examples of Heat-Treatable Wrought Alloys

12.4.1 Extruded Alloys

Following the fundamental work on the age-hardening responses in dilute Mg–Ca–Al and Mg–Ca–Zn alloys, Nakata et al. demonstrated that the Mg–Ca–Al–Mn-based alloys are extrudable at speeds as high as those used in 6XXX aluminum alloys (Nakata et al. 2015). As shown in the snapshots of Mg–0.3Al–0.15Ca–0.2Mn (AXM0301502) and AZ31 alloys extruded at 60 m/min (Fig. 12.12), the AXM0301502 extruded alloy shows a crack-free surface even at a die-exit speed of 60 m/min, while many cracks appear perpendicular to the extrusion direction on the surface of the AZ31 extruded alloy.

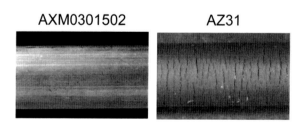

Fig. 12.12 Surface appearance of Mg–0.3Al–0.15Ca–0.2Mn (AXM0301502) and AZ31 alloys extruded at 60 m/min. Reprinted from ref. Nakata et al. (2015), Copyright 2012, with permission from Elsevier

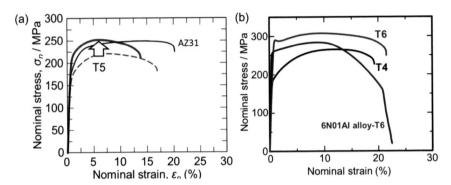

Fig. 12.13 Nominal stress–strain curves for AZ31 (**a**) and as-extruded and T5 treated AXM0301502 alloys. Note that the T5 treatment stands for the artificial aging to the peak hardness right after the extrusion. Nominal stress–strain curves for a T6-treated 6N01 Al alloy (**b**) and solution treated and T6-treated Mg–1.3Al–0.33Ca–0.46Mn (AXM10304) alloys. Reprinted from ref. Nakata et al. (2015, 2017a), Copyright 2012, with permission from Elsevier

The artificial aging for only 3 h immediately after extrusion leads to an increase in the yield strength from 170 to 207 MPa, which is higher than that of the AZ31 alloy due to the precipitation of the G.P. zones (Fig. 12.13 (a)). The yield strength was further increased to 290 MPa by composition optimization into Mg–1.3Al–0.33Ca–0.46Mn (wt.%) (AXM10304) and subsequent T6 treatment consisting of solution treatment and artificial aging (Fig. 12.13 (b)). Although the yield strength of 290 MPa is higher than that of the 6N01 aluminum alloy, such high strength is achieved at the expense of extrudability. The AXM10304 alloy is extrudable at 24 m/min, which is slower than the extrusion speed of AXM0301502 alloy, 60 m/min. However, Nakata et al. demonstrated that the excellent extrudability and the high yield strength can be simultaneously achieved by the optimization of the homogenization process prior to the extrusion; a Mg–1.6Al–0.2Ca–0.9Mn alloy can be extrudable at 60 m/min by and exhibits a high yield strength of 284 MPa (Nakata et al. 2017b).

Along with good extrudability and high strength, the dilute Mg–Ca–Al alloys also show an interesting feature. As shown in Fig. 12.13 (b), the dilute Mg–Ca–Al alloy does not show significant loss of ductility by the T6-treatment unlike precipitation hardened Mg–Al, Mg–Zn, and Mg–Sn alloys; the peak aged AXM10304 alloy shows a large elongation of 20%, which is comparable with that of the solution-treated samples.

12.4.2 Sheet Alloys

The large tensile elongation of the Mg–Ca–Al–Mn alloy indicates that the Mg–Al–Ca-based dilute alloy might also exhibit excellent room temperature formability and

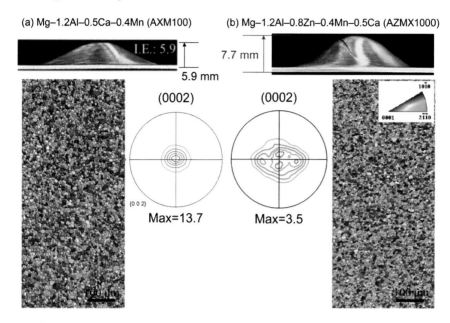

Fig. 12.14 Snapshot of the sample after the Erichsen cupping test, inverse pole figure map and (0002) pole figures of the solution treated (**a**) Mg–1.2Al–0.5Ca–0.4Mn (AXM100) and (**b**) Mg–1.2Al–0.8Zn–0.4Mn–0.5Ca (AZMX1000) alloys. Reprinted from ref. Bian et al. (2017), Copyright 2012, with permission from Elsevier

satisfactory strength (Bian et al. 2017). However, the solution treated Mg–1.2Al–0.5Ca–0.4Mn (AXM100) alloy sheet shows an I.E. value of only 5.9 mm, which is lower compared with that of the 6XXX series aluminum alloy, Fig. 12.14 (a). An inverse pole figure (IPF) map and (0002) pole figure (Fig. 12.14 (a)) obtained by electron backscatter diffraction (EBSD) indicate that the lack of formability is due to the development of a strong basal texture as discussed in Sect. 12.2.2 (Bian et al. 2016). Bian et al. reported that the poor room temperature formability of the Mg–0.5Ca–0.1Zr (at.%) alloy can be substantially improved by the addition of Zn (Bian et al. 2016). The Mg–1.2Al–0.8Zn–0.4Mn–0.5Ca (AZMX1000) alloy exhibited a high I.E. value of 7.7 in the solution treated condition since the addition of Zn caused the development of a "quadrupole" texture, in which (0002) poles were tilted toward RD and TD (Fig. 12.14 (b)).

The T6-treatment gives rise to the strength of the AZMX1000 alloy sheet (Bian et al. 2017). The AZMX1000 alloy reaches the peak hardness with only 1 h of artificial aging at 200 °C, and the yield strength of the AZMX1000 alloy increases from 177 to 204 MPa (Fig. 12.15 (a)).

In the actual car manufacturing process, body panels made from steel and aluminum alloys are heat treated at low temperatures for short time, e.g. at 170 °C for 20 min for paint curing. The phenomenon that the sheet materials are strengthened during the paint baking process is referred to as "bake-hardening." The bake-

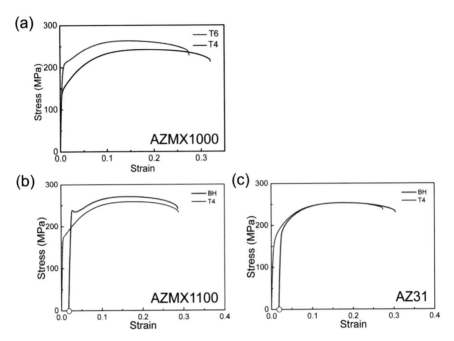

Fig. 12.15 (**a**) Nominal stress–strain curves for T4- and T6-treated Mg–1.2Al–0.8Zn–0.4Mn–0.5Ca (AZMX1000) alloys and nominal stress–strain curves for (**b**) Mg–1.3Al–0.8Zn–0.7Mn–0.5Ca (wt.%) (AZMX1110) and (**c**) AZ31 alloy before and after inducing 2% strain and artificial aging at 170 °C for 20 min. Reprinted from ref. Bian et al. (2017, 2018a), Copyright 2012, with permission from Elsevier

hardening process corresponds to strain aging, and the descriptions of the bake-hardenability are usually based on uniaxial tensile tests. First, 2% strain is induced into a specimen followed by artificial aging at 170 °C for 20 min. Note that the 2% strain is a typical amount of strain induced into the body materials during stamping, and artificial aging at 170 °C for 20 min simulates the industrial paint-baking process. The pre-strained and artificially aged sheet sample is again subjected to a tensile test. The bake-hardenability is defined as the difference between the maximum stress during pre-straining and the yield stress after baking.

The bake-hardenability in the magnesium alloy was demonstrated for the first time in the Mg–Ca–Al–Zn dilute alloy (Bian et al. 2018a). The AZMX1110 alloy exhibits a strength increase of approximately 40 MPa because of the bake-hardening process, while the commercial AZ31 alloy exhibits softening rather than strengthening, Fig. 12.15 (b) and (c) (Bian et al. 2018a). Consequently, the yield strength of the AZMX1110 alloy increases from 177 to 238 MPa through the bake-hardening process, which is significantly higher than that of the AZ31 alloy (186 MPa) ((Fig. 12.15 (c))).

Figure 12.16 shows the relationship between the yield strength and the I.E. values for various magnesium-based sheet alloys. The AZMX1110 alloy showed much

Fig. 12.16 Yield strengths and Index Erichsen values for various magnesium-based alloys. Reprinted from ref. Bian et al. (2018a), Copyright 2012, with permission from Elsevier

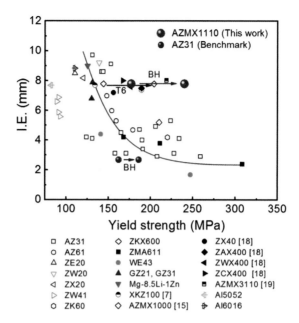

higher room temperature formability and yield strength compared with the existing alloys. Therefore, the bake-hardenable AZMX1110 sheet alloy successfully overcame the usual trade-off relationship between formability and strength, which has been hindering the application of magnesium sheet alloy (Bian et al. 2018a).

12.4.3 Toward the Improvement of Room Temperature Formability

The detailed analysis of the microstructure evolution during the stretch forming indicates the direction for the microstructure design to further improve the room temperature formability (Bian et al. 2017, 2018a). Figure 12.17 (a) and (b) shows IPF maps and (0002) pole figures of the AZMX1000 alloy stretch formed to dome heights of 3 and 7.7 mm during the Erichsen cupping test. The (0002) pole figures show that the angular distribution of these (0002) poles gradually becomes narrower toward the ND compared with those of the unstretched sample (Figs. 12.14 (b) and 12.17), indicating that the basal slip mainly operates during the stretch forming.

An EBSD-assisted slip trace analysis also revealed that the activation of the non-basal slip is the key to achieve excellent room temperature formability (Bian et al. 2018b). Note that the EBSD-assisted slip trace analysis initially uses EBSD orientation data to predict the traces generated by the major slip modes such as basal < a >, prismatic < a >, and pyramidal II < c + a >. Then, it compares the predicted traces with the traces detected from the SEM micrograph. The traces observed from the

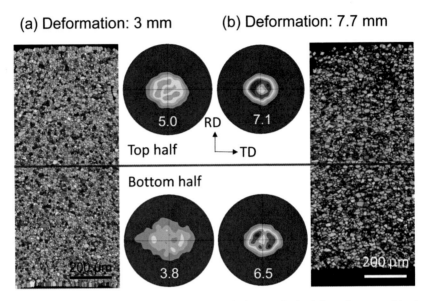

Fig. 12.17 Inverse pole figure maps and (0002) pole figures obtained from the top and bottom surfaces of the samples after (**a**) 3 mm and (**b**) 7.7 mm deformation by Erichsen cupping test. Reprinted from ref. Bian et al. (2017), Copyright 2012, with permission from Elsevier

SEM micrograph (Fig. 12.18 (a)) match well with the predicted trace of the basal < a > slip, indicating that these traces were generated by the basal < a > slip. The analytical results shown in Fig. 12.18 (b) indicate that the basal < a > slip is the dominant

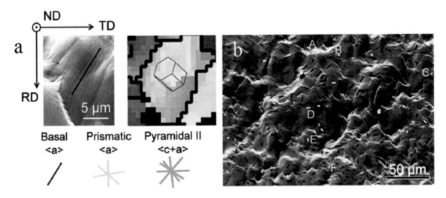

Fig. 12.18 (**a**) An SEM image and corresponding EBSD IPF map along with slip traces predictions for basal < a >, prismatic < a > and pyramidal ll < c + a > showing the methodology utilized for slip trace analysis on a stretch deformed grain. (**b**) SEM micrographs showing traces generated by basal < a >, prismatic < a > and pyramidal II < c + a > slips on the surface of the 3 mm stretch formed on Mg–1.2Al–0.3Ca–0.4Mn–0.3Zn alloy. Reprinted from ref. Bian et al. (2018b), Copyright 2012, with permission from Elsevier

deformation mode during the stretch forming of these samples, but Fig. 12.18 also shows the active operation of the prismatic < a > slip during the stretch forming.

Since the active operation of the prismatic < a > slip was also reported in fine-grained AZ31 and AZ61 alloys during the tensile test by Koike et al. (Koike et al. 2003; Koike and Ohyama 2005), the operation of the prismatic < a > slip would be a distinct feature in a magnesium alloy with relatively fine-grained structure. Zheng et al. also reported the enhanced work hardening and large elongation by the operation of < c + a > dislocations in an ultrafine-grained Mg–6.2%Zn–0.5%Zr–0.2%Ca (wt.%) alloy subjected to high-pressure torsion (HPT) (Zheng et al. 2019). Recent studies also reported that pure magnesium and magnesium alloys can show super-formability even at room temperature by grain boundary sliding and dynamic recrystallization when the average grain size is down to the micron scale (Somekawa et al. 2017, 2018; Zeng et al. 2017). Therefore, the formation of the fine-grained structure would be a key to further improve the room temperature formability.

12.4.4 Strengthening by G.P. Zones

The cause of the bake-hardenability of the AZMX1110 alloy sheet was investigated through a correlative analysis by TEM and 3D atom probe (3DAP). As shown in Fig. 12.19 (a) and (b), only a uniform imaging contrast can be observed in the BF-TEM image and selected area diffraction patterns obtained from the AZMX1110 alloy, which was artificially aged at 170 °C for 20 min after inducing 2% strain. Although it is difficult to understand the reasons for the bake hardening from the result of TEM analysis, a correlative TEM-3DAP analysis of the bake-hardened

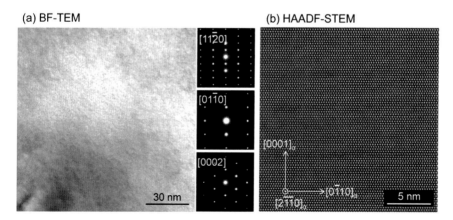

Fig. 12.19 (**a**) A bright-field TEM image and selected area diffraction patterns taken from the zone axes of $[11\bar{2}0], [01\bar{1}0]$, and $[0001]$. (**b**) HAADF-STEM image taken from Mg–1.3Al–0.8Zn–0.7Mn–0.5Ca (wt.%) (AZMX1110) alloy after inducing 2% strain and artificial aging at 170 °C for 20 min. Reprinted from Bian et al. (2018a), Copyright 2012, with permission from Elsevier

sample revealed the microstructure features that explain such reasons. Figure 12.20 (a) ~ (c) shows a two-beam BF-TEM image of the 3DAP specimen, the resulting 3D atom maps for the Al, Zn, Mn, and Ca, and the 3DAP map overlaid on the corresponding TEM image obtained from the bake-hardened sample, respectively. Four dislocations (designated as a, b, c, and d) are readily visible in the BF-TEM image (Fig. 12.20 (a)). They are confirmed to be segregated with Ca in the bake-hardened sample (Fig. 12.20 (b)). The 3DAP map overlaid on the corresponding TEM image in Fig. 12.20 (c) clearly shows this feature. The proxigram analyzed from the dislocation indicated by a red arrow shows that the concentrations of Al, Zn, and Ca in the dislocation core are significantly higher than those in the matrix (Fig. 12.20 (d)), which means that Ca, Al, and Zn are segregated along the dislocation cores. A closer look into the 3D atom map in Fig. 12.20 (e) also shows the presence of solute clusters enriched in Ca, Al, and Zn within the matrix. Therefore, Al, Zn,

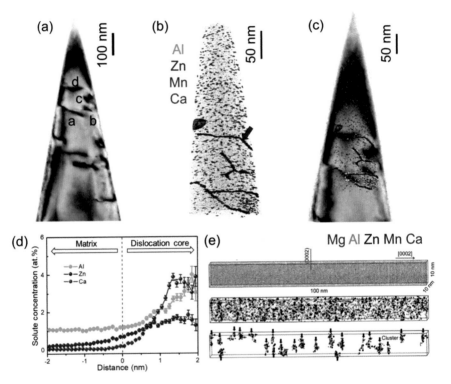

Fig. 12.20 (**a**) Bright-field image of a needle-shaped sample before 3DAP analysis, (**b**) 3D atom map of Mg, Ca, Zn, Al, and Mn, and (**c**) overlaid bright-field TEM and 3D atom map obtained from Mg–1.3Al–0.8Zn–0.7Mn–0.5Ca (wt.%) (AZMX1110) alloy after inducing 2% strain and artificial aging at 170 °C for 20 min. (**d**) is proxigrams of local solute concentrations at dislocation lines analyzed from the dislocation indicated by red arrow in (**b**) and (**e**) is 3DAP elemental mappings of Mg, solute atoms, and detected Al, Zn, and Ca clusters in the selected region. Reprinted from Bian et al. (2018a), Copyright 2012, with permission from Elsevier

and Ca atoms segregated into dislocation cores are considered to contribute to the bake-hardening effect along with co-clustering of these atoms.

12.5   Summary and Future Outlooks

In this chapter, a heat treatable alloy was introduced as a novel wrought magnesium alloy that would broaden alloy applications in the transportation sector. As a result of the development of precipitation hardenable magnesium alloys, which show substantial age hardening, Mg–Ca alloy micro-alloyed with Zn and/or Al was found to be promising as an industrially viable precipitation hardenable alloy. Mg–Al–Ca–Mn dilute alloy is extrudable at 60 m/min, which is comparable with the extrusion speed of the 6XXX series medium strength aluminum alloys, and the T6-treated sample exhibits higher yield strength (280 MPa) due to the dispersion of G.P. zones. From the Mg–Ca–Al–Zn alloy, a bake-hardenable alloy with excellent room temperature formability was also developed. This is the first time that bake hardenability was observed in magnesium alloys. The bake-hardenable Mg–Al–Ca–Zn–Mn alloy shows a stretch formability of 7.8 mm in I.E. value, which is comparable with that of the 6XXX series aluminum alloys. Bake hardening resulted in the segregation of solute elements along with the dislocation core and the solute cluster formation, and these effects increased the strength to 240 MPa. Although age-hardening phenomena have not been used in conventional wrought magnesium alloys, the optimized alloy shows sufficient age hardening response as observed in many heat-treatable aluminum alloys. Further development of heat-treatable magnesium alloys may reveal wider applications of wrought magnesium alloys.

Acknowledgements This work was supported by the Elements Strategy Initiative for Structural Materials (ESISM) of MEXT (grant number JPMXP0112101000), JSPS KAKENHI [grant number JP18H01756], and Advanced Low Carbon Technology Research and Development Program (ALCA) [grant number 12102886].

References

Atwell DL, Barnett MR (2007) Extrusion limits of magnesium alloys. Metall Mater Trans A 38:3032–3041. https://doi.org/10.1007/s11661-007-9323-2

Bhattacharjee T, Suh B-C, Sasaki TT et al (2014) High strength and formable Mg–6.2Zn–0.5Zr–0.2Ca alloy sheet processed by twin roll casting. Mater Sci Eng A 609:154–160. https://doi.org/10.1016/j.msea.2014.04.058

Bian MZ, Zeng ZR, Xu SW et al (2016) Improving formability of Mg-Ca-Zr sheet alloy by microalloying of Zn. Adv Eng Mater 18:1763–1769. https://doi.org/10.1002/adem.201600293

Bian MZ, Sasaki TT, Suh BC et al (2017) A heat-treatable Mg-Al-Ca-Mn-Zn sheet alloy with good room temperature formability. Scr Mater 138:151–155. https://doi.org/10.1016/j.scriptamat.2017.05.034

Bian MZ, Sasaki TT, Nakata T et al (2018a) Bake-hardenable Mg–Al–Zn–Mn–Ca sheet alloy processed by twin-roll casting. Acta Mater 158:278–288. https://doi.org/10.1016/j.actamat.2018. 07.057

Bian MZ, Sasaki TT, Nakata T et al (2018b) Effects of rolling conditions on the microstructure and mechanical properties in a Mg–Al–Ca–Mn–Zn alloy sheet. Mater Sci Eng A 730:147–154. https://doi.org/10.1016/J.MSEA.2018.05.065

Chino Y, Mabuchi M (2009) Enhanced stretch formability of Mg-Al-Zn alloy sheets rolled at high temperature (723 K). Scr Mater 60:447–450. https://doi.org/10.1016/j.scriptamat.2008.11.029

Chino Y, Sassa K, Mabuchi M (2008) Tensile Properties and Stretch Formability of Mg-1.5 mass%-0.2 mass%Ce Sheet Rolled at 723 K. Mater Trans 49:1710–1712. https://doi.org/10.2320/matertrans.MEP2008136

Chino Y, Sassa K, Mabuchi M (2009) Texture and stretch formability of a rolled Mg-Zn alloy containing dilute content of Y. Mater Sci Eng A 513–514:394–400. https://doi.org/10.1016/j.msea.2009.01.074

Chino Y, Huang X, Suzuki K et al (2010a) Influence of Zn concentration on stretch formability at room temperature of Mg–Zn–Ce alloy. Mater Sci Eng A 528:566–572. https://doi.org/10.1016/j.msea.2010.09.081

Chino Y, Huang X, Suzuki K, Mabuchi M (2010b) Enhancement of stretch formability at room temperature by addition of Ca in Mg-Zn alloy. Mater Trans 51:818–821. https://doi.org/10.2320/matertrans.M2009385

Chino Y, Ueda T, Otomatsu Y et al (2011) Effects of Ca on tensile properties and stretch formability at room temperature in Mg-Zn and Mg-Al alloys. Mater Trans 52:1477–1482. https://doi.org/10.2320/matertrans.M2011048

Elsayed FR, Sasaki TT, Mendis CL et al (2013) Significant enhancement of the age-hardening response in Mg–10Sn–3Al–1Zn alloy by Na microalloying. Scr Mater 68:797–800. https://doi.org/10.1016/j.scriptamat.2013.01.032

Huang X, Suzuki K, Saito N (2009) Textures and stretch formability of Mg-6Al-1Zn magnesium alloy sheets rolled at high temperatures up to 793 K. Scr Mater 60:651–654. https://doi.org/10.1016/j.scriptamat.2008.12.035

Huang X, Suzuki K, Chino Y, Mabuchi M (2011) Improvement of stretch formability of Mg-3Al-1Zn alloy sheet by high temperature rolling at finishing pass. J Alloy Compd 509:7579–7584. https://doi.org/10.1016/j.jallcom.2011.04.132

Huang X, Suzuki K, Chino Y (2012) Static recrystallization and mechanical properties of Mg–4Y–3RE magnesium alloy sheet processed by differential speed rolling at 823K. Mater Sci Eng A 538:281–287. https://doi.org/10.1016/j.msea.2012.01.044

Huang X, Suzuki K, Chino Y, Mabuchi M (2015a) Texture and stretch formability of AZ61 and AM60 magnesium alloy sheets processed by high temperature rolling. J Alloy Compd 632:94–102. https://doi.org/10.1016/j.jallcom.2015.01.148

Huang X, Suzuki K, Chino Y, Mabuchi M (2015b) Influence of aluminum content on the texture and sheet formability of AM series magnesium alloys. Mater Sci Eng A 633:144–153. https://doi.org/10.1016/j.msea.2015.03.018

Jayaraj J, Mendis CL, Ohkubo T, Hono K (2010) Enhanced precipitation hardening of Mg–Ca alloy by Al addition. Scr Mater 63:831–834. https://doi.org/10.1016/j.scriptamat.2010.06.028

Davis JR (ed) (1993) Aluminum and Aluminum Alloys. ASM International, Materials Park, OH

Kamado S, Ohara H, Kojima Y (eds) (2005) Advanced manufacturing technologies of magnesium alloy. CMC Publishing CO., LTD., Tokyo

Koike J, Ohyama R (2005) Geometrical criterion for the activation of prismatic slip in AZ61 Mg alloy sheets deformed at room temperature. Acta Mater 53:1963–1972. https://doi.org/10.1016/j.actamat.2005.01.008

Koike J, Kobayashi T, Mukai T et al (2003) The activity of non-basal slip systems and dynamic recovery at room temperature in fine-grained AZ31B magnesium alloys. Acta Mater 51:2055–2065. https://doi.org/10.1016/S1359-6454(03)00005-3

Abedesian M, Baker H (eds) (1999) Magnesium and magnesium alloys. ASM International, Materials Park, OH

Mackenzie L, Pekguleryuz M (2008) The recrystallization and texture of magnesium–zinc–cerium alloys. Scr Mater 59:665–668. https://doi.org/10.1016/j.scriptamat.2008.05.021

Mendis CL, Oh-ishi K, Hono K (2007) Enhanced age hardening in a Mg–2.4 at.% Zn alloy by trace additions of Ag and Ca. Scr Mater 57:485–488. https://doi.org/10.1016/j.scriptamat.2007.05.031

Murai T (2004) Extrusion of magnesium alloys. J Japan Inst Light Met 54:472–477

Nakata T, Mezaki T, Ajima R et al (2015) High-speed extrusion of heat-treatable Mg–Al–Ca–Mn dilute alloy. Scr Mater 101:28–31. https://doi.org/10.1016/j.scriptamat.2015.01.010

Nakata T, Xu C, Ajima R et al (2017a) Strong and ductile age-hardening Mg-Al-Ca-Mn alloy that can be extruded as fast as aluminum alloys. Acta Mater 130:261–270. https://doi.org/10.1016/j.actamat.2017.03.046

Nakata T, Xu C, Sasaki TT et al (2017b) Development of high-strength high-speed-extrudable Mg–Al–Ca–Mn Alloy. Magnes Technol 2017:17–21. https://doi.org/10.1007/978-3-319-52392-7

Nie J-F (2003) Effects of precipitate shape and orientation on dispersion strengthening in magnesium alloys. Scr Mater 48:1009–1015. https://doi.org/10.1016/S1359-6462(02)00497-9

Nie J-F (2012) Precipitation and hardening in magnesium alloys. Metall Mater Trans A 43:3891–3939. https://doi.org/10.1007/s11661-012-1217-2

Nie J-F, Gao X, Zhu SM (2005) Enhanced age hardening response and creep resistance of Mg-Gd alloys containing Zn. Scr Mater 53:1049–1053. https://doi.org/10.1016/j.scriptamat.2005.07.004

Oh J, Ohkubo T, Mukai T, Hono K (2005) TEM and 3DAP characterization of an age-hardened Mg–Ca–Zn alloy. Scr Mater 53:675–679. https://doi.org/10.1016/j.scriptamat.2005.05.030

Oh-ishi K, Watanabe R, Mendis CL, Hono K (2009) Age-hardening response of Mg–0.3 at.% Ca alloys with different Zn contents. Mater Sci Eng A 526:177–184. https://doi.org/10.1016/j.msea.2009.07.027

Park SJ, Jung HC, Shin KS (2017) Deformation behaviors of twin roll cast Mg-Zn-X-Ca alloys for enhanced room-temperature formability. Mater Sci Eng A 679:329–339. https://doi.org/10.1016/j.msea.2016.10.046

Ringer SP, Hono K, Sakurai T (1995) The effect of trace additions of Sn on precipitation in Al-Cu alloys: An atom probe field ion microscopy study. Metall Mater Trans A 26:2207–2217. https://doi.org/10.1007/BF02671236

Sasaki TT, Oh-ishi K, Ohkubo T, Hono K (2006) Enhanced age hardening response by the addition of Zn in Mg–Sn alloys. Scr Mater 55:251–254. https://doi.org/10.1016/j.scriptamat.2006.04.005

Sasaki TT, Elsayed FR, Nakata T et al (2015) Strong and ductile heat-treatable Mg–Sn–Zn–Al wrought alloys. Acta Mater 99:176–186. https://doi.org/10.1016/j.actamat.2015.06.060

Somekawa H, Kinoshita A, Kato A (2017) Great room temperature stretch formability of fine-grained Mg-Mn alloy. Mater Sci Eng A 697:217–223. https://doi.org/10.1016/j.msea.2017.05.012

Somekawa H, Singh A, Sahara R, Inoue T (2018) Excellent room temperature deformability in high strain rate regimes of magnesium alloy. Sci Rep 8:1–9. https://doi.org/10.1038/s41598-017-19124-w

Suzuki K, Chino Y, Huang X et al (2015) Enhanced room-temperature stretch formability of Mg 0.2 mass % Ce alloy sheets processed by combination of high-temperature pre-annealing and warm rolling. Mater Trans 56:1096–1101. https://doi.org/10.2320/matertrans.L-M2015811

Trang TTT, Zhang JH, Kim JH et al (2018) Designing a magnesium alloy with high strength and high formability. Nat Commun 9:2522. https://doi.org/10.1038/s41467-018-04981-4

Wu D, Chen RS, Han EH (2011) Excellent room-temperature ductility and formability of rolled Mg-Gd-Zn alloy sheets. J Alloy Compd 509:2856–2863. https://doi.org/10.1016/j.jallcom.2010.11.141

Yi S, Park JH, Letzig D et al (2016) Microstructure and mechanical properties of Ca containing AZX310 alloy sheets produced via win toll casting technology. Magnes Technol 2016:389–393

Zeng Z, Nie JF, Xu SW et al (2017) Super-formable pure magnesium at room temperature. Nat Commun 8:1–5. https://doi.org/10.1038/s41467-017-01330-9

Zheng R, Bhattacharjee T, Gao S et al (2019) Change of deformation mechanisms leading to high strength and large ductility in Mg-Zn-Zr-Ca alloy with fully recrystallized ultrafine grained microstructures. Sci Rep 9:11702. https://doi.org/10.1038/s41598-019-48271-5

Printed in the United States
by Baker & Taylor Publisher Services